The Soviet Union's Invisible Weapons of Mass Destruction

"We are not playing here, we are making a weapon!"

—Major General Nikolai Nikolaevich Urakov, Director of *Biopreparat*'s Institute of Applied Microbiology, Obolensk

"We planned to introduce new properties into disease organisms, such as antibiotic resistance, altered antigen structure and enhanced stability in the aerosol form, making delivery of the agent easier and more effective....We also reasoned that the genes for toxins or other factors could be introduced into the cells of bacteria or the DNA of viruses and thereby yield strains with wholly new and unexpected pathogenic properties."

—Igor' Valerianovich Domaradskii, Second Deputy Director of Science at *Biopreparat*'s Institute of Applied Microbiology

Anthony Rimmington

The Soviet Union's Invisible Weapons of Mass Destruction

Biopreparat's Covert Biological Warfare Programme

palgrave
macmillan

Anthony Rimmington
BIRMINGHAM, UK

ISBN 978-3-030-82881-3 ISBN 978-3-030-82882-0 (eBook)
https://doi.org/10.1007/978-3-030-82882-0

This Palgrave Macmillan imprint is published by the registered company Springer Nature Switzerland AG.
The registered company address is: Gewerbestrasse 11, 6330 Cham, Switzerland

To
Igor' Valerianovich Domaradskii,
Vladimir Artemovich Pasechnik
and
Kanatzhan Baizakovich Alibekov
who shone a light on the heart of darkness

Acknowledgements

The author would like to pay tribute to the high-level *Biopreparat* scientists who hosted himself and colleagues at their facilities at a time when their world had been ripped asunder, with the USSR lying in ruins and political and economic chaos everywhere apparent. The first of these was Major General (Reserves) Nikolai Nikolaevich Urakov, who invited the author and his colleagues to visit *Biopreparat*'s applied microbiology institute in Obolensk on a number of occasions in the early 1990s. Some in the international disarmament community mockingly referred to him as "Darth Vader", after the villainous *Star Wars* character. However, underneath a mask of stern rectitude, there was a genuine human side to his character and, after a few meetings, he was revealed to possess a strong sense of humour. Another scientist with a military background was Colonel (Reserves) Gennadii Nikolaevich Lepeshkin. In July 1995, he hosted the author, who was part of a UK group visiting the former *SNOPB* facilities, which had been formerly focused on the large-scale production of weaponized anthrax bacteria, in Stepnogorsk. Even in very difficult economic circumstances, he did his utmost to make his guests feel comfortable. At one stage, he very kindly invited the UK party to visit his apartment in Stepnogorsk town centre where he allowed them to view his Soviet army uniform and medals. He was very convivial company and was enormously proud of his facility's achievements, both during the Soviet period and in newly independent Kazakhstan. Finally, there was Vladimir Petrovich Zav'yalov, who on more than one occasion during the 1990s, hosted UK visitors to *Biopreparat*'s Institute of Immunology in Lyubuchany. He revealed himself to be a scientist of exceptional calibre with an array of international publications. More than that, he has since proved himself to be an individual with an uncompromising adherence to strong moral and ethical principles and values and was understandably well-liked in the wider Russian scientific community.

It is necessary to acknowledge the author's wider debt to the former Soviet scientists, who, at considerable personal risk, have provided accounts of their participation in the covert activities of *Biopreparat*. These include, especially, Igor' Valerianovich Domaradskii, Vladimir Artemovich Pasechnik and Kanatzhan Baizakovich Alibekov. Without them, it would not have been possible to complete this current study and historians of the Soviet Union will be forever in their debt. One should note also the enormous contribution to our understanding of the Soviet offensive BW programme made, through a series of publications, by the Russian chemist, Lev Aleksandrovich Fedorov. Mention should also be made of Valerii Pavlovich Mityushev who has shone new light on the origins of the *Biopreparat* network.

The author owes a great debt of gratitude to the monumental scholarship undertaken over an 11-year period by Milton Leitenberg and Raymond Zilinskas with Jens Kuhn. This resulted in the publication in 2012 of their authoritative work of reference, *The Soviet Biological Weapons Programme: A History*. Sadly, Zilinskas, one of the world's leading authorities on the Soviet BW effort, died after a brief illness on 14 September 2018.

Alex Donaldson very kindly read and commented on the manuscript as it was being finally prepared. A debt of thanks is owed to Dr Christopher Joyce, who, as a postgraduate student during the period 1998–2005, assisted the author with accessing documentation from the Russian State Archive of the Economy (*Rossiiskii gosudarstvennyi arkhiv ekonomiki* or *RGAE*) and the Russian State Library, both based in Moscow. The author also wishes to acknowledge the kind donation made to him by Reg Bottomley, Director of Peruvum Ltd., of his personal archive of company documentation relating to business visits to life sciences organizations in the Soviet Union, undertaken during the period 1977–1985. Reg, now deceased, probably had the distinction of being among the very few Westerners to become personally acquainted with Major General (Reserves) Vsevolod Ivanovich Ogarkov, who had served as the first head of *Biopreparat*.

Contents

About the Author

Anthony Rimmington is a former senior research fellow at the University of Birmingham's Centre for Russian, European and Eurasian Studies (CREES). He was the winner, as a postgraduate student at this institution, of the John Grayson Memorial Prize. He has published widely on the civil life sciences industry in Russia and the former Soviet Republics and is the author of *Technology and Transition: A Survey of Biotechnology in Russia, Ukraine and the Baltic States* (1992). Other publications on this topic include: *Tekhnologiya i perekhodnyi period. Obzor biotekhnologii v Rossii, na Ukraine i v stranakh baltii*, in *Biotekhnologiya, ekologiya, meditsina: Materialy III–IV Mezhdunarodnykh nauchnykh seminarov 2001–2002*, Volga-Vyatka Centre of Applied Microbiology, Kirov, 2002, pp. 75–78; Biotechnology Legislation in Central & Eastern Europe, European Federation of Biotechnology Task Group on Public Perceptions of Biotechnology's *Briefing Paper No. 9*, June 1999, p. 4; Biotechnology and Industrial Microbiology Regulations in Russia and the Former Soviet Republics, in Hambleton, P., Melling, J., Salusbury, T.T. (Eds.), *Biosafety in Industrial Biotechnology*, Blackie Academic & Professional, London, 1994, pp. 67–89; Perestroika and Soviet Biotechnology, *Technology Analysis and Strategic Management*, Vol. 2, No. 1, 1990, pp. 63–79; Soviet Biotechnology: The Case of Single Cell Protein, in Amann, R., Cooper, J. (Eds.), *Technical Progress and Soviet Economic Development*, Basil Blackwell, Oxford, 1986, pp. 75–93; and Issues in Soviet Biotechnology: The Case of Single-Cell Protein, in *Adaptability to New Technologies of the USSR and East European Countries*, NATO Colloquium, Brussels, 17–19 April, 1985, pp. 217–234.

Numerous contributions regarding the Soviet and Russian civil life sciences industry have also been made to a number of popular scientific publications including *New Scientist*, *Bio/Technology*, *International Industrial Biotechnology*, *The Genetic Engineer and Biotechnologist*, *European Microbiology* **and** *Microbiology Europe*.

Rimmington has also written extensively on the Soviet Union's offensive biological weapons programme including *Stalin's Secret Weapon: The Origins of Soviet Biological Warfare* (2018) and *The Soviet Union's Agricultural Biowarfare Programme: Ploughshares to Swords* (Palgrave Macmillan, 2021). He has completed a series of journal articles and book chapters on the subject including: From Offence to Defence? Russia's Reform of Its Biological Weapons Complex and the Implications for Western Security, *The Journal of Slavic Military Studies*, Vol. 16, No. 1, 2003, pp. 1–43; The Soviet Union's Offensive Programme: The Implications for Contemporary Arms Control, in Wright, S. (Ed.), *Biological Warfare and Disarmament: New Problems/New Perspectives*, Rowman and Littlefield Publishers, Inc., Lanham, 2002, pp. 103–150; Invisible Weapons of Mass Destruction: The Soviet Union's BW Programme and Its Implications for Contemporary Arms Control, *The Journal of Slavic Military Studies*, Vol. 13, No. 3, September 2000, pp. 1–46; Konversion Sowjetischer BW-produktions-einrichtungen: Der fall *Biomedpreparat*, Stepnogorsk, Kasachstan, in Buder, E. (Ed.), *Möglichkeiten und Grenzen der Konversion von B-Waffen-Einrichtungen*, Lit Verlag, Münster, 2000, pp. 245–262; Fragmentation and Proliferation? The Fate of the Soviet Union's Offensive Biological Weapons Programme, *Contemporary Security Policy*, Vol. 20, No. 1, April 1999, pp. 86–110; Conversion of BW Facilities in Kazakhstan, in Geissler, E., Gazsó, L., Buder, E. (Eds.), *Conversion of Former BTW Facilities*, NATO Science Series, Kluwer Academic Publishers, London, 1998, pp. 167–186; and From Military to Industrial Complex? The Conversion of Biological Weapons' Facilities in the Russian Federation, *Contemporary Security Policy*, Vol. 17, No. 1, April 1996, pp. 80–112.

During the last years of the Cold War, Rimmington travelled to several research establishments at sites located across the Soviet Union. Following the collapse of the USSR, he was allowed access to a number of former Soviet life sciences R&D institutes and manufacturing facilities with a view to assisting with their participation in international non-proliferation programmes.

List of Figures

CHAPTER 1

Introduction

This is the first academic study to be completed which focuses solely on *Biopreparat*, the Soviet agency created in April 1974, which spearheaded the largest and most sophisticated BW programme the world has ever seen. At its height *Biopreparat* employed 30–40,000 personnel and incorporated five major military-focused research institutes, numerous design and instrument-making facilities, three pilot plants and five dual-use production plants. The network pursued major offensive R&D programmes, such as *Bonfire* and *Factor*, and genetically engineered microbial strains which were resistant to an array of antibiotics. In addition, using these techniques, bacterial agents were created with the ability to produce various peptides, yielding strains with wholly new and unexpected pathogenic properties. It was only in 1988, after the defection to the UK of a key scientist, Vladimir Artemovich Pasechnik, and the implementation of economic and political reforms by the Soviet leadership, that *Biopreparat* slowly began to emerge also, as the most important player in the USSR's civil biopharmaceutical sector. This process was given fresh impetus in December 1991 with the collapse of the Soviet Union and the placing of *Biopreparat* under Russian control. The US subsequently played a key role in assisting Russia with the demilitarization of the network and the destruction of dual use BW facilities.

It is not only the scale of the *Biopreparat* programme which makes it so globally significant but also the apparent ease with which it enabled the

A. Rimmington, *The Soviet Union's Invisible Weapons of Mass Destruction*, https://doi.org/10.1007/978-3-030-82882-0_1

Soviet Union to circumvent its obligations under the terms of the 1972 Biological and Toxin Weapons Convention (BTWC). The offensive BW effort, subject to extraordinary levels of secrecy, was effectively concealed within the USSR's civil microbiological industry, and with which it was intimately entwined. Elaborate cover stories were generated and adhered to, and, despite their best endeavours, Western intelligence analysts struggled to clearly understand what was going on. It was only through a lucky break, the defection of the high-level *Biopreparat* scientist, Pasechnik, to the UK, that the full scale and extent of the programme was eventually understood. All this serves to highlight a crucial flaw in the BTWC, the lack of provisions to ensure the compliance of nations that have signed and ratified it.

An array of newly available sources has been consulted for this project. For the first time in the West, the author has succeeded in gaining access to copies of official Soviet documents which are held within the Russian State Archive of the Economy (*Rossiiskii gosudarstvennyi arkhiv ekonomiki* or *RGAE*). These provide unique insight into the development of *Biopreparat* and its transformation from a military-focused Soviet science production association into a Russian open joint stock company with ambitious commercial aspirations. The author has also been able to gain access to visit reports and correspondence relating to the interaction of a number of UK companies with *Biopreparat* and its affiliated facilities. This includes business correspondence with Kanatzhan Baizakovich Alibekov—who was a key source of information on the *Biopreparat* programme for Western intelligence. Finally, this study is also the first to present analysis of newly released Russian memoirs written by Valerii Pavlovich Mityushev who had intimate knowledge of *Biopreparat*'s network of dual-use BW production plants and its crucially important design institute *VNIIbiokhimmashproekt*.

This present study places *Biopreparat* firmly in its Soviet historical context. It draws on the newly available sources, to argue that scholars should consider shifting the whole historical timeframe with regard to Soviet BW activity. For the first time, it is argued that the Soviet Union's network of dual-use BW production facilities had in fact been established at least a decade prior to the creation of *Biopreparat*, under the leadership of Nikita Sergeyevich Khrushchev. At the very heart of this earlier effort was a new, previously unknown, military unit, the Scientific-Research Technical Bureau (*Nauchno-issledovatel'skoe tekhnicheskoe byuro* or *NITB*). *NITB* was attached to the USSR Ministry of Defence's Fifteenth Administration, headed by Colonel-General Efim Ivanovich Smirnov, which had overall charge of the Soviet Unions' BW programme. The bureau had been created in August 1958, in accordance with a resolution of the CPSU and the USSR Council of Ministers. Its prime function was to create covert dual-use BW facilities at a number of pharmaceutical and microbiological enterprises. Over the next decade or so, dual-use BW production plants were created at Berdsk, Omutninsk, Penza and Kurgan. It is apparent from this evidence, that previous perceptions by Western scholars of the Khrushchev era as contributing little to the development of the Soviet Union's biological warfare capabilities are incorrect. This was in fact a pivotal period in the Soviet programme, when BW production technology was being transferred from the military to facilities covertly concealed within civil manufacturing plants.[1] This was later to manifest itself as a key feature of the subsequent *Biopreparat* programme.

Another organization which was in existence nearly a decade prior to the formation of *Biopreparat* was the Main Administration of the Microbiological Industry under the USSR Council of Ministers (*Glavnoe upravlenie mikrobiologicheskoi promyshlennosti pri Sovete Ministrov SSSR* or *Glavmikrobioprom*). This study reveals that Vasilii Dmitrievich Belyaev, the individual placed in charge of the microbiological industry after its creation in February 1966, was a former head of a dual-purpose chemical facility called Factory No. 91 located at Beketovka to the south of Stalingrad. In 1960, he had been awarded the Soviet Union's most prestigious honour, the Lenin Prize, for achieving full-scale production at the plant of the nerve agent Sarin. Belyaev's experience of successfully managing one of the most secret of the Soviet Union's dual-use chemical plants made him an ideal choice for heading *Glavmikrobioprom* which had itself absorbed the USSR's dual-use biological factories and begun to run its own BW research projects. Later, in the 1970s, Belyaev was then placed in ultimate charge of the All-Union Science Production Association *Biopreparat*, which was created as a subordinate unit under *Glavmikrobioprom*. Belyaev's key role in the management of the Soviet offensive BW programme is fully documented in this study for the first time.

Another major new finding in this study concerns the central role of the All-Union Scientific-Research Design Institute of Applied Biochemistry (*Vsesoyuznyi nauchno-issledovatel'skii proektno-konstruktorskii institut prikladnoi biokhimii* or *VNIIbiokhimmashproekt*) within *Biopreparat*. No less than nine senior personnel were transferred from the USSR Ministry of Defence's *NITB* (see above), a military unit focused on the creation and maintenance of the Soviet Union's dual-use BW production plants, to *VNIIbiokhimmashproekt*. The bulk of the remaining high-level staff at the new institute also had military backgrounds. Such was the institute's importance that it initially occupied rooms within *Biopreparat*'s headquarters building. It went on to design *Biopreparat*'s major BW R&D centres at Obolensk and Kol'tsovo. *VNIIbiokhimmashproekt* also played a key role in the engineering and maintenance of the Berdsk, Omutninsk and Stepnogorsk dual-use BW production plants.

The military accomplishments of the *Biopreparat* programme are documented in the text. These briefly are reported to include, as part of the secret programme *Bonfire*, the genetic engineering of a range of bacteria including a strain of *Bacillus anthracis* (the causative agent of anthrax) which was made resistant to seven or eight antibiotics. In another innovative project, *Francisella tularensis* (the causative agent of tularaemia) bacteria were apparently coated with Protein A to allow them to bypass the immunodefence system in humans. In another genetic engineering project, the antigenic presentation of the surface of an F1-minus strain of *Yersinia pestis* (the causative agent of plague) was altered. It was anticipated that this would lead to delays in detecting and identifying the pathogen in any countries attacked with this strain. *Biopreparat* scientists are also reported to have developed technology for the production of a dry formulation of Marburg virus which could be easily dispersed in an aerosol. In perhaps, the most concerning of the programmes, *Factor*, researchers successfully created genetically engineered bacteria with the ability to produce various peptides which yielded strains with wholly new and unexpected properties.[2]

One of the key features of the historical BW programmes pursued by the Soviet Union had been the occurrence of accidents in the wake of sometimes slipshod safety procedures. The most notorious of these incidents concerned the USSR Ministry of Defence's Scientific-Research Institute of Bacterial Vaccine Preparations, part of Military Compound Number 19 in Sverdlovsk. On Monday, 2 April 1979, an anthrax aerosol was released from this facility which led directly to the deaths of at least 68 people in Sverdlovsk itself and to cases of animal anthrax in nearby villages (Rudnii, Bol'shoe Sedelnikovo, Maloe Sedelnikovo, Pervomaiskii, Kashino and Abramovo) to the south-east of the city. It was the first major indication in the West that the Soviet Union had embarked upon an offensive biological weapons effort.[3] *Biopreparat* proved to be no exception with regard to the occurrence of major incidents, the most serious of which occurred in mid-April 1988 at its virology institute in Kol'tsovo during the pursuit of a Marburg virus weaponization project. During the course of an experimental procedure performed by an inexperienced colleague, Nikolai Vasil'evich Ustinov was stabbed in the finger with the needle of a syringe being used to extract blood from a guinea pig which had been previously injected with Marburg virus. Ustinov died on 30 April in terrible agony, bleeding profusely and unable to move or even open his eyes. During the period through to his death, Ustinov maintained a log in which he recorded his feelings and symptoms and which he hoped would lead to an improved understanding of the virus and its effects. *Biopreparat* scientists subsequently isolated a new Marburg virus variant, named Variant U, from samples of Ustinov's organs and this was absorbed into the institute's culture collection.[4]

On a more positive note, the history of *Biopreparat* encompasses two of the most significant achievements with regard to biological nonproliferation programmes implemented in Russia and Kazakhstan after the end of the Cold War. The first of these concerns US cooperation with Russia with regard to the *Sibbiofarm* Production Association (Berdsk). Here, the US' CRDF Global assisted with training, procurement of equipment and renovation of the plant. A key part of the interaction also concerned the removal and destruction of its dual-use BW facilities including the dismantling of several large fermenters. *Sibbiofarm* was later judged by CRDF Global to be both profitable and sustainable and the facility has now emerged as one of Russia's largest producers of commercial biologics. The outcome can be considered a major success for both parties. From Russia's point of view, the success of the project means that decades of significant Soviet investment have not been simply thrown away and the revitalized facility can now be considered a major growth pole for the Novosibirsk region. For the US, the project has demonstrated both, to the Russian government, and the local scientific community, that it is a reliable partner and will pursue agreed collaborative projects through to their successful conclusion. In addition, it has succeeded in eliminating significant dual-use biological facilities which had been in existence for more than half a century.

The second project concerns US cooperation with Kazakhstan concerning the dismantlement of former *Biopreparat* facilities at Stepnogorsk. These had been utilized for the large-scale production of weaponized *B. anthracis* (the causative agent of anthrax) with a view to its potential use as an offensive agent. In 1996, Kazakhstan agreed to the dismantlement of the plant in cooperation with the US Department of Defence. This involved the complete demolishment of production, drying and indoor testing and nutrient medium facilities at the site. Again, this project can be considered a major success for all parties. The authorities in Kazakhstan recognized that they simply lacked the requisite finance and relevant expertise to attempt any wholesale conversion of the remote site. Clearly, such was the scale of the weapons complex, that even advanced Western nations would have found such a project a formidable and enormously expensive challenge. So, for them, complete elimination of BW facilities can be considered a highly desirable outcome. For the US too, the project can be considered as a major achievement—it had succeeded in eliminating the most advanced and sophisticated BW production complex the world has ever seen.

Biopreparat's switch in focus from military to civil objectives was not just a result of interactions with US non-proliferation agencies but resulted also from shifting priorities in the political leadership. During the 1970s and 1980s, *Biopreparat* had already demonstrated its potential capabilities with regard to healthcare projects, with the initiation of the large-scale industrial production of a new influenza vaccine. Production eventually reached 19.9 million doses by 1986. The election of Mikhail Gorbachev in March 1985 as General Secretary of the CPSU gave added impetus to this shift towards the pursuit of civil projects. In November of that same year, *Biopreparat* was incorporated into a new super-ministry which aimed to use its expertise in cutting-edge technologies in molecular biology—which were beginning to have such an impact in the West—for production of a new generation of medicines. While there was no substantive diminishment in *Biopreparat*'s offensive BW programmes at this time, the political leadership were clearly setting major pointers to the future direction of travel.

The role of individual Soviet scientists with regard to the curtailment of *Biopreparat*'s military programmes should not be underestimated. In October 1989, Vladimir Artemovich Pasechnik, Director of one of the agency's most important R&D facilities, the All-Union Scientific-Research Institute of Highly Pure Biopreparations (Leningrad), defected to the UK. In his subsequent debriefings he revealed an astonishing amount of detail regarding *Biopreparat*'s network and its enormous offensive BW programme. The damage inflicted on *Biopreparat*'s military programmes by the Soviet whistle-blower cannot be overstated. According to one of those who interviewed him, "*The fact that Vladimir defected was one of the key acts of the entire ending of the Soviet Union and the end of the Cold War. It was the greatest breakthrough we ever had*".[5] There were other influential Soviet scientists too, who were seeking to redirect *Biopreparat* towards healthcare rather than warfare. Chief among them was Kanatzhan Baizakovich Alibekov, First Deputy Chairman of the Board of the reorganized *Biopreparat*, who began in 1990 to explore the opportunities for engagement with Western pharmaceutical companies including the possibility of organizing the joint production of genetically engineered human insulin. He also led negotiations with a UK company, Oxford Virology, with regard to the formation of a partnership for the international export of *Biopreparat*'s products.

In April 1991, *Biopreparat* was again reorganized, absorbing an array of new R&D and production facilities, and finally emerging as the most powerful organization in the Soviet Union with regard to domestic production of antibiotics, vaccines and other essential medicines. The collapse of the USSR occurred shortly thereafter, in December 1991, leaving *Biopreparat* to bestride the emerging Russian biopharmaceutical sector like a colossus. While there remained powerful voices supporting the retention of its military capabilities, they were to be subsequently swept aside, as a result of both intense Western diplomatic pressure and the frantic power struggle in Russia to secure control of *Biopreparat*'s enormously valuable commercial assets.

In the new period of building capitalism, local entrepreneurs initially focused on the more lucrative economic sectors of the Russian Federation, such as banking and exportable raw materials and succeeded in gaining control of ministries with responsibility for these sectors. Attention then shifted to remaining opportunities in less substantive markets such as the Russian pharmaceutical industry. Vacroux argues, "*Placing their representatives near or at the top of important federal and regional bureaucracies allowed these entrepreneurs to capture some of the key regulators of their industry and to lobby the remaining parts of the government for benefits 'from within'*".[6] Despite its military connections, *Biopreparat* was not spared attention from such entrepreneurs and there were intense struggles for control of the most lucrative parts of the network.

What of the fate of *Biopreparat* in present-day Russia? Certainly, it now appears to be the case that if it exists at all, then it is confined to the shadows and no longer takes central stage. In the decade following the collapse of the Soviet Union, the profitable parts of its empire were privatized and became firmly focused on commercial projects. Other facilities have been transferred to federal control. The State Scientific Centre of Virology and Biotechnology *Vektor* (Kol'tsovo), formerly subordinate to *Biopreparat*, is one such example. It is now subordinate to the Federal Service for Surveillance on Consumer Rights Protection and Human Wellbeing and has emerged as one of Russia's leading centres for disease surveillance and research. In addition, former dual-use weapons facilities, such as those at *Sibbiofarm*, have been eliminated. It is apparent then, that *Biopreparat*'s main raison d'être, the management of a network of R&D and production facilities has all but been taken away. Most of the key personnel who participated in Soviet programmes are in any case deceased or have taken up much more lucrative positions in the commercial pharmaceutical sector. Military biological programmes are once again the sole preserve of the Russian Ministry of Defence. Whether *Biopreparat* exists in some form or not, perhaps then becomes a moot point. It may simply be the case that it finds itself to be increasingly irrelevant and to have no useful role to play in Russia's new, post-Soviet landscape.

Notes

1. In August 1958, Khrushchev also launched an offensive agricultural BW programme embracing six institutes and placed it under the control of a secret department within the USSR Ministry of Agriculture. See Rimmington, *The Soviet Union's Agricultural Biowarfare Programme: Ploughshares to Swords*, Palgrave Macmillan, 2021.
2. The achievements of the *Biopreparat* programme with regard to the weaponization of bacterial and viral pathogens is fully documented in Leitenberg, M., Zilinskas, R.A., with Kuhn, J.H., *The Soviet Biological Weapons Programme: A History*, Harvard University Press, Cambridge, Massachusetts, 2012.
3. Rimmington, A., *Stalin's Secret Weapon: The Origins of Soviet Biological Warfare*, Hurst & Company, London, 2018, p. 23.
4. Raskryty smerty uchenykh-virusologov v tsentre "Vektor", Angarskii Portal Novostei, 19 February 2020, https://xn%2D%2D38-6kcaak9aj5chl4a3g.xn%2D%2Dp1ai/raskryty-smerti-uchenyh-virusologov-v-centre-vektor/.
5. Hoffman, D.E., *The Dead Hand: The Untold Story of the Cold War Arms Race and Its Dangerous Legacy*, Doubleday, London, 2009, p. 332.
6. Vacroux, A., Regulation and Corruption in Transition: The Case of the Russian Pharmaceutical Markets, pp. 133–151 in Kornai, J., Rose-Ackerman, S., *Building a Trustworthy State in Post-Socialist Transition*, Palgrave Macmillan, New York, 2004.

CHAPTER 2

The Secret History: Khrushchev's Creation of Soviet Reserve Biological Warfare Mobilization Facilities Within Civil Production Plants in the 1950s and 1960s

Up until now scholarly analysis of the Soviet BW programme has presumed, in the absence of any evidence to the contrary, that the creation of biological warfare mobilization facilities within civil production facilities was initiated in the mid-1970s by *Biopreparat*. However, significant new evidence has emerged which points strongly to the fact that the USSR, under the leadership of Nikita Sergeyevich Khrushchev, had begun to create this network at least a decade earlier than had been previously assumed. At the very heart of this effort was a new Scientific-Research Technical Bureau (*Nauchno-issledovatel'skoe tekhnicheskoe byuro* or *NITB*). *NITB* was attached to the USSR Ministry of Defence's Fifteenth Administration, headed by Colonel-General Efim Ivanovich Smirnov, which managed the Soviet Union's BW programme.[1] The bureau had been created in August 1958, in accordance with a resolution of the CPSU and the USSR Council of Ministers. It was initially located within a military base located on ulitsa Matrosskaya Tishina (literally meaning "Seaman's Silence") in the Sokol'niki region in Moscow. This street is the site of an infamous Russian prison which dates back to the 1870s and prior to that an insane asylum which dates back to the eighteenth century. Later, at the end of 1960, *NITB* was transferred to Moscow's Lefortovo district. The new bureau was charged with the task of developing technology for use in the creation of dual-use pharmaceutical and microbiological enterprises. Mobilization capacity would be put in place which in the event of an outbreak of

A. Rimmington, *The Soviet Union's Invisible Weapons of Mass Destruction*, https://doi.org/10.1007/978-3-030-82882-0_2

hostilities would have the capability of producing "*special products*" (i.e. biological weapons).[2]

The first major project to be a focus for *NITB* concerned a manufacturing facility for microbiological products, nestled beside the river Berd in the small town of Berdsk, around 26 km to the south of Novosibirsk. The creation of the new plant, named the Berdsk Chemical Factory (*Berdskii khimicheskii zavod*), was initiated in response to a decree on the development of the Soviet chemical and microbiological industry issued on 25 February 1955 by the USSR Council of Ministers.[3] It was one of the dozens of new chemical and biological facilities created at around this time in the Soviet Union. Actual construction of the Berdsk plant began in 1957 but only at a snail's pace. It was only after the holding of a Plenum of the Central Committee of the CPSU in December 1963, which sharply criticized the pace at which construction of chemical plants was taking place across the country, that real progress was made on the Berdsk project, and the first finished product, an antibiotic used in animal feed, was produced in facility No. 32.

Another key event at the plant was the appointment in 1963 of 37-year-old Yurii Nikolaevich Timon'kin as director. He was born on 9 January 1926 in Leningrad and lived in the city until he was 15. After the outbreak of the Second World War, he and his mother were evacuated to Berdsk. His father, Nikolai Nikolaevich, a journalist, was left behind in Leningrad and perished in the besieged city in 1942. Timon'kin was drafted into the army in 1945 and served for five years as a driver in Siberia. Returning to Berdsk he became seriously involved in sports and emerged as a regional swimming champion. During the period 1950–1955, he studied at the Higher Party School in Novosibirsk and then returned to Berdsk where he served as head of the Department of Industry and Transport at the city party committee. He then took up the position of secretary of the party committee at the Berdsk Radio Plant which had 15,000 employees on its staff.[4] In 1963 he was appointed director of the Berdsk Chemical Factory and served in this position until 1970. Timon'kin clearly lacked any obvious qualifications with regard to the management of a microbiological plant, but the authorities recognized that he was a natural leader and a decision-maker and after his appointment, he was able to rapidly create a professional, efficient team around him and infect them with his enthusiasm. However, all was not entirely in harmony and during this period Timon'kin's crucial relationship with the plant's chief engineer, Gennady Egorovich Skvortsov, became an increasingly difficult one. It was against this background that, in 1970, Timon'kin grasped the opportunity offered to him by the authorities to take up the post of director of a major new microbiological production facility being constructed in Stepnogorsk, Kazakhstan (see below).[5] He was replaced as director at Berdsk in 1973 by Boris Vasil'evich Prilepskii, originally a graduate from the Frunze Higher Naval School in Leningrad.[6]

By 1965, the Berdsk Chemical Factory had initiated the manufacture in facility No. 33 of two microbial pesticides, Entobacterin and Dendrobacillin, both based on the bacterium *Bacillus thuringiensis*.[7] Interestingly, when the UK's David Kelly arrived at the plant nearly three decades later, in October 1993, as part of a joint UK/US inspection team, he was informed that the plant's four 64,000-litre fermenters which were then in use, were turning-out *B. thuringiensis*-based pesticides.[8] The Berdsk Chemical Factory was officially subordinate to the Novosibirsk regional economic council (*sovnarkhoz*). Sovnarkhozes were introduced by Nikita Khrushchev in May 1957 in what was to be a short-lived and failed attempt to combat the centralisation and departmentalism of the national branch ministries.[9] After the liquidation of the sovnarkhoz, the plant was in a state of limbo for around a year and then transferred to the USSR Ministry of Agriculture, which had no real interest in managing the facility.[10] The plant was presumably then transferred to the microbiological industry in 1966, prior to being attached to *Biopreparat* in 1974.

According to the newly released memoirs of Valerii Pavlovich Mityushev, who was employed as a Lieutenant within *NITB*, the Berdsk Chemical Factory was, from its very inception, dual-use in nature.[11] As part of *NITB*'s role in creating mobilization capacity at the site, Mityushev began to make working visits to the plant in 1965, which corresponded to the period when control automation systems were being introduced, the first of their kind in this branch in the Soviet Union. He was to continue to make visits related to the factory's biological weapons capacity over the next two decades. Mityushev does not give any information relating to the BW capability created at the site. Fortunately, in his own memoirs regarding his participation in the *Biopreparat* programme, Kanatzhan Baizakovich Alibekov, writing as Ken Alibek, does provide some details, noting, "*The* (Berdsk) *facility, built in the 1960s, had been used mainly as a reserve plant for assembling and filling bomblets with weaponised bacteria that would be produced at its own and other installations*".[12]

A second major focus for the USSR Ministry of Defence's Scientific-Research Technical Bureau was another manufacturing plant, the

Omutninsk Chemical Factory (*Omutninskii khimicheskoi zavod* or *OKhZ*). The factory was created in accordance with a decree issued on 2 August 1958 by the CPSU and the USSR Council of Ministers. On 12 December 1958, the Executive Committee of the Kirov Regional Council of Workers' Deputies selected a remote site in an area of swamps and forests, for the construction of *OKhZ*, 3 km outside the settlement of Vostochnyi (named after Yurii Gagarin's *Vostok* spacecraft). The settlement is located close to Omutninsk, around 150 km to the north-east of Kirov. Building work at Vostochnyi began a short time later towards the end of 1961.[13] Both the accommodation for workers and the plant itself were built contemporaneously. The construction work was undertaken by convict labour teams provided by the Ministry of Internal Affairs, who were supervised by guards with dogs. One of the convict brigades, comprised of electricians, was led by a former chief power engineer at a large factory in Gorky. This energetic individual had received a ten-year sentence after crushing a man to death whilst driving under the influence of alcohol. The presence of the convicts at Vostochnyi had a negative influence on the settlement and there were a number of incidents involving fighting and even on occasion, murder.[14]

The construction work at Vostochnyi did not proceed well and this situation was not helped by constant changes in the senior management. The plant's first director, K.I. Platonov, had been appointed on 15 July 1961. A short time later, on 1 November 1964, Platonov was replaced by M.V. Demidov. However, it was the appointment on 22 January 1968 of a young, dynamic new director, Vladimir Arkad'evich Valov, that rapidly changed the situation on the ground at Vostochnyi. The 35-year-old Valov, described as chubby, with blonde hair and ruddy cheeks, had quickly risen to prominence at the local branch of the CPSU in Kirovo-Chepetsk where he had served since 1965 as Chairman of the City Executive Committee.[15] Having surveyed the situation upon his arrival in Vostochnyi, Valov immediately took the decision to suspend the ongoing construction of the plant and to focus all the material and human resources which were available to him, on the development of the infrastructure of the settlement itself. This encompassed the construction of additional accommodation, a good school, a club house and a hockey court—with Valov himself, an enthusiast of the sport, taking a lead role in the organization of local hockey teams. The Vostochnyi hockey team would go on to be champions of the Kirov region on more than one occasion. Once Valov had created a substantive reserve of new apartments at Vostochnyi, he used these, together with promises of promotions, to lure to the site well-qualified specialists from nearby industrial centres such as Gorky and Dzerzhinsk (see later). It was at this point, with the requisite number of specialists in place, that Valov resumed construction of *OKhZ*.[16] The first branch of *OKhZ*, comprising Buildings Nos. 1 and 2, an administration building and a range of ancillary services, was brought on stream in December 1968 and the first product, a veterinary antibiotic, was produced in 1969.[17] Large-scale fermenters, $15m^3$ and $63m^3$ in size, were employed in the manufacturing process.

During the Soviet period, in "closed" or sensitive settlements such as Vostochnyi, along with the regular authorities, there also existed a system of special police, prosecutors and courts. It is interesting that the authorities in Moscow chose to appoint Ivan Lavrinenko as special prosecutor in Vostochnyi. For he had previously worked as an assistant prosecutor in the ultra-secret closed nuclear city known as Krasnoyarsk-26 (Zheleznogorsk). It had been established in 1950 for the production of weapons-grade plutonium. Lavrinenko's appointment is certainly indicative of the high-level defence status of the Omutninsk Chemical Factory which was viewed by the authorities as being on a par with that of the nuclear facilities belonging to the USSR Ministry of Medium Machine Building (*Ministerstvo srednego mashinostroeniya SSSR* or *Minsredmash*).[18]

The dual-use nature of the Omutninsk Chemical Factory, from its very inception, is again pointed to in Mityushev's memoirs (see above). Over a period of a quarter of a century, in connection with the creation and maintenance of its mobilization capacity, he was a regular visitor to the plant. During the first ten-year period this was as a representative of the USSR Ministry of Defence.[19] There was, initially at least, no outward sign of the special defence status of the Omutninsk Chemical Factory. It was officially subordinate to the Administration of the Metallurgical and Metalworking Industry (*Upravleniya metallurgicheskoi i metallo-obrabatyvayushchei promyshlennosti*) of the Kirov regional economic council (*sovnarkhoz*). Like its sister facility in Berdsk, *OKhZ* was presumably absorbed in 1966 by the newly created microbiological industry prior to being transferred to *Biopreparat* in 1974.

There were in addition at least two more Soviet production plants in which reserve mobilization capacity was at this time installed by the USSR Ministry of Defence's Scientific-Research Technical Bureau. Mityushev, reports in his memoirs, that from 1965 through to 1971, he had been a frequent visitor to the Penza Factory of Medical Preparations (*Penzenskii zavod meditsinskikh preparatov*)—later to take the name *Biosintez* (*Biosynthesis*).[20] Penza is located some 625 kilometres south-east of Moscow on the Sura River and was originally founded as a frontier fortress city. The plant was created in 1955, presumably as a result of the decree issued on 25 February and referred to above. It was located on the eastern outskirts of the city, around 25 minutes from the city centre. By October 1959, the plant had produced its first products—the antibiotic, biomycin and vitamin B12.[21] The factory was to go on to become one of the most important producers of antibiotic substances in the Soviet Union.[22] Fedorov notes that this plant "***had the built-in ability to instantly switch over to biological warfare agents production***".[23] *NITB* also possessed mobilization capacity at the Penza Factory's sister facility, the Kurgan Factory of Medical Preparations (*Kurganskii zavod meditsinskikh preparatov*)—later, in 1973 taking the name *Sintez* (*Synthesis*)—which was located in another old fortress town in the Urals, Kurgan. Engineers and specialists for the new plant were recruited from the Leningrad Chemical-Pharmaceutical Institute and the Moscow Institute of Fine Chemical Technology. Construction of the plant began in 1956 with convict labour provided by the Ministry of Internal Affairs again providing a critical input. The heart of the facility was Building 101 which housed 32 fermenters—16 in each wing of the building, each with a capacity of 50 cubic metres.[24] Construction work was finally completed at the beginning of September 1958 and the veterinary antibiotic biomycin produced during the first fermentation process. In February 1959, vitamin B12 was manufactured and later that year other products followed, such as microbial biomass for use as animal feed and gibberellin, a plant growth stimulator.[25] Later, there was more of a focus on medicines for human use, and in August 1966, production of the antibiotic oxytetracycline began and in 1969, phenoxymethyl-penicillin, another antibiotic, was manufactured.[26]

One interesting aspect of the reserve BW network created by Khrushchev is that the *NITB* placed one of its own staff as "*curator*" on-site at each plant where it had created mobilization capacity. The curator at Penza, for example, was Mikhail Sergeevich Molchanov. Meanwhile, *NITB*'s curator at Kurgan was a local man, Yurii Dmitrievich Zotov, a graduate of a Soviet military chemical academy.[27] The presence on the ground of these military personnel, who did not have any formal position within the management hierarchy of the production plants, was a vital guarantor that installed mobilization capacities were maintained in a state of immediate readiness should they be required. Interestingly, *Biopreparat* was later to mimic this system and appoint its own curators to facilities where the main emphasis was on civil, rather than military programmes (see below).

It was to be around a decade after their construction was initiated, that Western intelligence, employing signal intelligence and imagery, began to identify the key participants in the clandestine Soviet BW mobilization network. Koblentz reports that in 1971, based on an analysis of satellite photographs, the CIA first identified the Omutninsk Chemical Factory as a suspected BW production facility. While the CIA continued to have doubts as to how far this facility was involved in the USSR's BW programme, the State Department's Bureau of Intelligence and Research was convinced of its role. In a comprehensive assessment written in 1975, Gary Crocker, a CBW analyst for the department, determined that the Omutninsk plant had the same munition storage bunkers as those identified at military BW facilities in Zagorsk and Sverdlovsk and "*was also implicated in military activities based on intercepts collected by the National Security Agency*".[28] Likewise, by 1976, the CIA had evidence pointing to suspected biological warfare production and storage at Berdsk.[29] However, it was not until 1984 that the mobilization facilities at Penza and Kurgan were identified in US press reports along with four other suspected sites.[30] Koblentz notes that the most suspicious indicators of BW activities at all of the suspect Soviet sites was the existence of bunkers which *"suggested that the plants were involved not just in research but also in production and/or storage of biological weapons. The presence of identical configurations at civilian microbiological sites, such as Omutninsk, Pokrov and Berdsk also suggested a military purpose for these sites*". He also notes that communications intercepts were used by analysts to identify suspicious activities including the use of encrypted communications and/or the frequency of contacts with known military sites.[31]

One of the only references to the existence of the Soviet BW mobilization network to have been made in the Russian media appeared in an article in February 1998 in the newspaper *Moskovskaya pravda*. The author of the report, Lev Aleksandrovich Fedorov, an important whistle-blower in the chemical and biological weapons arena, noted the comments made by the local Vice Governor of the Penza *oblast'*, Yu.A. Laptev, on 31 October 1997. He stated, "*We should have had bacteriological weapons. In that industrial complex 'Biosintez' special sections and a huge medical unit were built, that's where there were a special laboratory, isolation blocks and so on, the fences were formidable...So, we were lucky we have nuclear and we have chemical weapons and we were on the point of having bacteriological ones*".[32]

It is now apparent that previous perceptions by Western scholars of the Khrushchev era as contributing little to the development of the Soviet Union's biological warfare capabilities are incorrect. This present study reveals that major BW mobilization facilities were at this time incorporated into four newly constructed microbiological and pharmaceutical production factories. In addition, a previously unidentified Ministry of Defence Scientific-Research Technical Bureau was created in August 1958 to manage the installation and maintenance of the BW mobilization capacity. Unofficial "curators" with no formal positions in the management hierarchy, were appointed by *NITB* to each of the plants to ensure that installed mobilization capacities were maintained in a state of immediate readiness should they be required. This was a pivotal moment in the Soviet programme, when BW production technology was being transferred from the military to facilities covertly concealed within civil manufacturing plants. This was later to manifest itself as a key feature of the *Biopreparat* programme. The author's previous study also reveals that it was exactly at this moment that Khrushchev launched a major offensive agricultural BW programme, embracing the construction of six specialized research institutes working on the development of anti-crop and anti-livestock weapons.[33] There is no doubt therefore that during this period, there was a significant acceleration of the Soviet BW programme and that Soviet officials working under Khrushchev played a crucial role in initiating the transfer of military BW production technologies to the civil sector.

Notes

1. Poklonskii, D.L., K 60-letiyu sozdaniya nauchno-issledovatel'skogo tekhnicheskogo byuro, *Vestnik voisk RKhB zashchity*, Vol. 2, No. 4, 2018, http://supotnitskiy.ru/vestnik_voysk_rchbz/vestnik_rchbz_8_9.pdf, Accessed on 17 July 2020.
2. Mityushev, *Zapiski obyknovennogo cheloveka*, http://www.mybio.ru/zapiski/text, Accessed on 17 July 2020, p. 347.
3. Sibbiofarm, Istoriya predpriyatiya, http://www.sibbio.ru/about/history/, Accessed on 20 July 2020.
4. Mityushev, *Zapiski obyknovennogo cheloveka*, pp. 533–534.
5. Ibid., p. 534.

6. Rimmington, A., *Who's Who in Russia & Republics Biotechnology*, Technology Detail, York, 1994, p. 233.
7. Sibbiofarm, Istoriya predpriyatiya, http://www.sibbio.ru/about/history/, Accessed on 20 July 2020.
8. Mangold, T, Goldberg, J., *Plague Wars: A True Story of Biological Warfare*, Macmillan, 1999, pp. 197–198.
9. For a full discussion of Khrushchev's failed reform, see Hanson, P., *The Rise and Fall of the Soviet Economy: An Economic History of the USSR from 1945*, Pearson Education, London, 2003, pp. 58–60.
10. Timon'kin, Yu.N., *Ya tvoeyu sud'boyu zhivu*, Akmola, 1992, https://kzref.org/ya-tvoeyu-sudeboyu-jivu.html?page=3, Accessed on 9 October 2020.
11. Mityushev, *Zapiski obyknovennogo cheloveka*, p. 532.
12. Alibek, K., with Handelman, S., *Biohazard: The True Story of the Largest Covert Biological Weapons Programme in the World – Told From the Inside by the Man Who Ran It*, Hutchinson, London, 1999, p. 41.
13. Vostok (zavod), *Vikipediya*, ru.wikipedia.org › wiki › Восток_Восток (завод), Accessed 17 July 2020.
14. Mityushev, *Zapiski obyknovennogo cheloveka*, pp. 521–522.
15. Valov Vladimir Arkad'evich, Geroi Sotsialisticheskogo truda, pochetnyi zhitel' pgt Vostochnyi, http://www.vostokuprava.ru/vostok/jitili/valov, Accessed on 22 July 2020.
16. Mityushev, *Zapiski obyknovennogo cheloveka*, p. 523.
17. Vostok (zavod), Vikipediya, ru.wikipedia.org › wiki › Восток_Восток (завод), Accessed 17 July 2020.
18. Mityushev, *Zapiski obyknovennogo cheloveka*, p. 529.
19. Ibid., p. 527.
20. Ibid., p. 515.
21. Biosintez, kombinat meditsinskikh preparatov, http://www.penza-trv.ru/go/region/biosintez, Accessed on 23 July 2020.
22. Istoriya predpriyatiya, http://biosintez.com/ru/about/history, Accessed 23 July 2020.
23. Fedorov, L.A., *Soviet Biological Weapons: History. Ecology. Politics*, Krasand, Moscow, 2013, p. 77.
24. Mityushev, *Zapiski obyknovennogo cheloveka*, p. 510.
25. Smi o nas, 15 September 2018, http://ksintez.ru/media/about-us/kurganskomu-zavodu-meditsinskikh-preparatov-sintez-60-let-/, Accessed on 23 July 2020.
26. Ibid.
27. Mityushev, *Zapiski obyknovennogo cheloveka*, pp. 510–515.
28. Koblentz, G.D., *Living Weapons: Biological Warfare and International Security*, Cornell University Press, London, 2011, p. 164.

29. Ibid.
30. Anderson, J., Van Atta, D., How Russia Fights with Poison and Plague, *Reader's Digest*, Vol. 125, No. 750, October 1984, p. 59.
31. Koblentz, *Living Weapons*, p. 163.
32. Fedorov, L., Seks bomba po sovetski: Gotova li Rossiya k biologicheskoi voine?, *Moskovskaya pravda*, 19 February 1998, p. 2.
33. Rimmington, A., *The Soviet Union's Agricultural Biowarfare Programme: Ploughshares to Swords*, Palgrave Macmillan, 2021.

CHAPTER 3

Glavmikrobioprom and the Emergence of the Soviet Microbiological Industry

In 1966, a major impetus was given to the development of the newly emerging industrial microbiology sector in the Soviet Union (see Fig. 3.1). On 18 February of that year, the USSR Council of Ministers issued a decree "***Concerning the development of the microbiological industry and about the organization of the management of this industry***".[1] Under the terms of this decree, the Main Administration of the Microbiological Industry under the USSR Council of Ministers (*Glavnoe upravlenie mikrobiologicheskoi promyshlennosti pri Sovete Ministrov SSSR* or *Glavmikrobioprom*) was created. Some 62 enterprises previously associated with timber processing, the food industry and agriculture were initially transferred to *Glavmikrobioprom*.[2] The latter was headquartered at ulitsa Lesteva, 18 in Moscow's Danilovsky District.[3] On 18 February 1970, Belyaev, the new head of the *glavk* (see below), wrote a letter to Leonid Brezhnev regarding the general results achieved by the microbiological industry, mainly focusing on several shortcomings in the industry especially in relation to agriculture.[4]

Very soon after its creation, *Glavmikrobioprom* was to absorb the most important Soviet R&D unit engaged in fundamental research in molecular genetics. The Sector of Genetics and Selection of Microorganisms had been organized within the Radiobiological Department at the USSR Academy of Sciences' Institute of Atomic Energy (Moscow) by Sos Isaakovich Alikhanyan. The institute had formerly been known as

A. Rimmington, *The Soviet Union's Invisible Weapons of Mass Destruction*, https://doi.org/10.1007/978-3-030-82882-0_3

Fig. 3.1 USSR Stamp to commemorate Ninth International Microbiology Congress, Moscow, July 1966 (USSR Poat—Scanned 600 dpi by User Matsievsky from personal collection, Public Domain, https://commons.wikimedia.org/w/index.php?curid=41031619, Accessed on the 28 July 2020)

Laboratory No. 2 and had taken a lead role in the Soviet military programme to develop the atomic bomb. Genetics and molecular biology research within the Institute of Atomic Energy were initiated under the terms of a resolution issued in August 1958 by the Central Committee of the Communist Party and the USSR Council of Ministers entitled "*About Atomic Technology-Oriented Research in the Fields of Biology and Radiobiology*". As a result of this resolution, a Radiobiological Department (*Radiobiologicheskii otdel* or *RBO*), headed by V. Yu. Gavrilov (formerly employed within the Arzamas-16 nuclear weapons production centre), was established at the atomic research institute.[5] The *RBO* and Alikhanyan's subordinate unit appear to have been created as part of an initiative by Academicians Igor' Vasil'evich Kurchatov and Anatolii Petrovich Aleksandrov to provide a refuge at the Institute of Atomic Energy for Soviet geneticists who were being persecuted in a virulent political campaign directed by Trofim Denisovich Lysenko. It was in the face of these political attacks that sometime around 1948, Alikhanyan—who himself spoke out bravely in defence of Mendelian genetics—had been forced to leave the Department of Genetics at Moscow State University. After his eventual transfer to the Kurchatov Institute, Alikhanyan initiated and led the first Soviet fundamental research in molecular genetics.

On 16 February 1968 *Glavmikrobioprom* issued an order under which Alikhanyan and his colleagues were transferred to a newly created institute under its auspices, the All-Union Scientific-Research Institute of Genetics and Selection of Industrial Microorganisms (*Vsesoyuznyi nauchno-issledovatel'skii institut genetiki i selektsii promyshlennykh mikroorganizmov* or *VNIIgenetika*). Alikhanyan was appointed director of the new facility and construction of a new complex to house the institute began immediately.[6] Alikhanyan appears to have made promises to Belyaev to resolve issues relating to biological weapons.[7] This may explain why *Glavmikrobioprom* acted so swiftly to establish the new research institute. Nothing is known about any military projects subsequently pursued by Alikhanyan, but he was given a place on the Interdepartmental Council for Molecular Biology and Genetics which later played a key role in the management of the Soviet offensive BW programme (see below).[8] Alikhanyan also led a major research programme at *VNIIgenetika* focused on the isolation of antibiotic-producing microbial strains and their subsequent application in industry. The new institute also played a crucial role in developing the Soviet Union's capabilities with regard to the production of restriction enzymes, which are a vital requirement for work in molecular biology (see below).

The case of *VNIIgenetika* illustrates the early connections which were formed between the Soviet atomic weapons complex and the offensive biological programmes. Management of the atomic bomb project, which included the Institute of Atomic Energy, was in fact the responsibility of the USSR Ministry of Medium Machine Building (*Minsredmash*) under Efim Pavlovich Slavskii.[9] Vasilii Dmitrievich Belyaev (head of *Glavmikrobioprom*—see below) and Slavskii are reported to have had very close personal relations with the latter, assisting in a number of *Glavmikrobioprom* projects including the construction of facilities in Stepnogorsk. It is possible that their friendship originated during discussions concerning the transfer of *VNIIgenetika* to Belyaev's *glavk*.

In addition to *VNIIgenetika*, the Soviet authorities also decided to create a powerful new R&D facility for the Soviet microbiological industry. On 18 February 1968, the CPSU Central Committee and USSR Council of Ministers issued Decree No. 102 on the creation of the All-Union Scientific-Research Institute of Microbial Pesticides and Bacterial Preparations (*Vsesoyuznyi nauchno-issledovatel'skii institut mikrobiologicheskikh sredstv zashchity rastenii i bakpreparatov* or *VNIIbakpreparat*) within *Glavmikrobioprom*. In 1974, Yurii Tikhonovich Kalinin, who later was to head *Biopreparat*, was appointed a deputy director of the institute. Prior to taking up this position, Kalinin had been serving as an aide to a member of *Glavmikrobioprom*'s Scientific-Technical Council, so would have had an intimate knowledge of the *glavk*'s R&D programmes. It was envisaged that a new building in Moscow to house the institute would be ready by 1978.[10] Later, in late June 1984, under Order No. 276, *VNIIbakpreparat* was renamed the All-Union Scientific-Research Institute of Biotechnology (*Vsesoyuznyi nauchno-issledovatel'skii institut biotekhnologii* or *VNIIbiotekhnologii*).[11] A branch of the institute was also established in Kazan', capital of the Republic of Tatarstan.

Another key R&D institute created by *Glavmikrobioprom* at this time was the Special Design Bureau of Biologically Active Substances (*Spetsial'noe konstruktorsko-tekhnologicheskoe byuru biologicheski aktivnykh veshchestv* or *SKTB BAV*). It was organized in accordance with an order from the USSR Council of Ministers in 1970 as a sub-division of the Berdsk Chemical Factory. It was located immediately next to the area occupied by the plant. In January 1971, it acquired the status of an independent entity and was no longer considered part of the chemical factory. Some 6.9 million roubles were earmarked for investment in the facility which comprised a laboratory and design building (4800 m^2), a vivarium for small animals (2000 m^2) and a building housing an experimental workshop (800 m^2). Apartments for staff members were obtained from the adjacent factory.[12] The facility had a pivotal new role in the Soviet microbiological industry, going on to supply more than 100 biochemical and microbiological reagents necessary for research in molecular biology. Immediately upon its creation, *Biopreparat* absorbed the bureau and on 27 November 1981 renamed it as the Scientific-Research Design Institute of Biologically Active Substances (*Nauchno-issledovatel'skii konstruktorsko-tekhnologicheskii institut biologicheski aktivnykh veshchestv* or *NIKTI BAV*).[13]

In a further attempt to strengthen the R&D base of the new *glavk*, especially with regard to its military capabilities, it was also considered desirable for it to acquire an institute focused on pathogens research. It was initially intended that the Volgograd branch of the Rostov-on-Don Scientific-Research Anti-Plague Institute would be transferred to *Glavmikrobioprom*. However, there was strong resistance to the planned move from Avetik Ignat'evich Burnazyan, a deputy minister of the USSR Ministry of Health. Burnazyan successfully argued that instead, *Glavmikrobioprom* should focus on the creation of its own dedicated network of R&D facilities, a recommendation which was subsequently to be vigorously pursued after the creation by the *glavk* of the *Biopreparat* network.[14]

The Soviet Microbial Proteins Programme

The primary task immediately assigned to *Glavmikrobioprom* (see Fig. 3.2) by Soviet planners concerned the production of microbial proteins (single-cell proteins or SCP) which could be used to eliminate the USSR's acute and growing protein deficit in livestock feed supplies. The lack of a protein supplement in Soviet feedstuffs was developing into a critical situation where only one pound of beef was being produced for every 15 pounds of feed. This compared to the situation on farms in the US, where, with the use of a 10–15 per cent soya meal supplement, one pound of beef was being produced for only 8–9 pounds of feed. It was therefore apparent that a large quantity of grains could be saved if the protein deficiency in

Fig. 3.2 Pavilion of the microbiological industry at the Exhibition of the Achievements of the National Economy (*VDNKh*) (AlixSaz, The main facade of the pavilion Microbiological industry (former Oilseeds): Mira str., house 119, building 30, Ostankino North East district of Moscow, Attribution-ShareAlike 4.0 International (CC BY-SA 4.0), Location: 55° 50′ 6.45″ N, 37° 37′ 8.08″ E, Created: 23 November 2016, https://commons.wikimedia.org/wiki/ File:The_main_facade_of_the_pavilion_Microbiological_industry.jpg, Accessed on the 18 August 2020)

Soviet supplies were to be made up. It was this realization which prompted the USSR Council of Ministers in 1960 to begin a search for new sources of protein which could be added to animal feeds. Against a background of intensive worldwide interest in SCP during the 1950s and 1960s, Soviet economic planners eventually came to the conclusion that large-scale industrial production of microbial proteins offered the best solution to the growing protein gap in animal feed production.[15] From the strategic point of view, it was also envisaged by the Soviet leadership that domestic production within the microbiological industry would allow the country to become self-sufficient in protein feed additives and reduce or eliminate the requirement for large-scale grain imports.

Microbial proteins had in fact first found widespread use in the USSR during the Second World War. In November 1941, during the long and enormously destructive siege of Leningrad by Nazi Germany's Army Group North, a team led by V.I. Shakova developed technology which used wood-shavings to produce SCP.[16] By the spring of 1942, several plants in the besieged city were producing SCP using this method and yeast steaks and yeast milk were then prepared from it for consumption in local restaurants.[17] Later, in Moscow from 1942 to 1943, and in the larger towns of the Urals and Siberia from 1943 to 1944, dozens of small yeast-producing plants were brought on stream.[18] Two of the first industrial-scale SCP production units began operation in 1943 at the Khor Hydrolysis Factory (Khabarovsk *krai*) and the Solikamsk Sulphite-Alcohol Factory (Perm' *oblast'*).[19] Thus, during this period, for the first time in the Soviet Union, SCP was perceived as constituting a potentially valuable source of protein. In addition, the military significance of such production became apparent with microbial proteins being reliably produced directly in cities, and not, as in the case of grain supplies, having to be transported from war-affected areas.

Glavmikrobioprom was to take a twin-track approach to the production of microbial proteins. The first strand of its programme was to focus on the production of SCP from cellulose using hydrolysis of either solid feedstocks, such as wood-waste or corn cobs, or a liquid feedstock such as the waste liquor generated during pulp and paper production. Dilute sulphuric acid was added to the selected feedstock and heated in a hydrolysis tank. The hydrolysis step produced simpler carbohydrates which could be more readily digested by yeast in a fermenter, after which the microbial protein was purified, dried and bagged for shipment.[20] Up until this point, the hydrolysis industry had focused on the production of ethanol, but this was no longer economically viable and a number of factors favoured a switch to large-scale production of microbial proteins. These included the fact that the microorganisms already in use, *Saccharomyces* and *Candida* were later shown to be ideal for the safe, rapid production of SCP and that "*it was a relatively easy job to convert the anaerobic alcohol processes to aerobic SCP processes*".[21]

During the formation of *Glavmikrobioprom*, it absorbed both a major R&D institute, the All-Union Scientific-Research Institute of Hydrolysis (*VNIIgidroliz*), together with the pre-existing plants belonging to the All-Union Hydrolysis Agency (*Gidrolizprom*). It then went on to convert existing hydrolysis factories and also proceeded with the construction of several purpose-built facilities, with production of cellulose-based microbial proteins eventually reaching 600,000 tonnes by 1988.[22] However, the cellulose-based production was soon to be eclipsed in scale and importance by an emerging new technology in the Soviet Union for the production of microbial proteins from petroleum derivatives.

"Food from Oil": Soviet Production of Microbial Proteins from Petroleum Derivatives

As indicated above, *Glavmikrobioprom* also pursued a second major strand of its microbial proteins programme which focused on production from liquid alkanes obtained from crude oil. Certainly, there was no lack of such feedstock and the crude oil found in the huge Volga-Ural fields in 1960 had a high n-alkane (waxy) content. In 1963, the CPSU Central Committee instructed the USSR State Committee for the Coordination of Scientific Research Work and the USSR Council of Ministers to examine the possibility of industrial production of microbial proteins from petroleum hydrocarbons. A special scientific council, headed by Academician Nikolai Dmitrievich Ierusalimskii, was created under the Presidium of the USSR Academy of Sciences as a result, to run a wide-ranging research and design programme. The latter focused on three main areas: research on the microbiology and technology for SCP production from n-alkanes, studies on the biological value and safety of microbial proteins when fed to agricultural animals; and studies to assess the biological value and safety to humans of livestock products derived from animals fed with microbial proteins.[23]

In 1963, the All-Union Scientific-Research Institute of the Biosynthesis of Protein Substances (*Vsesoyuznyi nauchno-issledovatel'skii institut biosinteza belkovykh veshchestv* or *VNIIsintezbelok*) was formed on the basis of the Moscow Division of the Scientific-Research Institute of the Hydrolysis and Sulphite-Alcohol Industry. It was assigned the task of obtaining microbial protein from petroleum.[24] *VNIIsintezbelok* was subsequently absorbed by the newly created *Glavmikrobioprom* and spearheaded the Soviet Union's petroleum-based microbial proteins production programme. Later, it would also play a small, but nonetheless significant role, in the Soviet Union's offensive BW programme.

By 1964, Soviet scientists had succeeded in creating a pilot plant in Krasnodar for the production of microbial protein—given the name protein-vitamin concentrate (*belkovo-vitaminnyi kontsentrat* or *BVK*)—from petroleum.[25] This had a capacity of 1500 tonnes per annum. Tests revealed, however, that the use of an impure gas oil substrate led to toxicological problems with residual hydrocarbons. In 1968, with *Glavmikrobioprom* now coordinating the programme, a second pilot plant with a capacity of 12,000 tonnes was built in Ufa. This time purified liquid alkanes were used as substrate and a wide range of biological tests appeared to demonstrate a lack of harmful side-effects when microbial protein from the plant was used in animal feedstuffs.[26]

It was in the wake of these successful projects that the 24th party congress of the CPSU outlined an ambitious new Five-Year Plan for the development of the microbiological industry during the period 1971–1975. It called, initially, for the construction of four very large-scale microbial protein plants utilizing purified n-alkanes as feedstock.[27] Later, during the period through to the collapse of the USSR, construction was ongoing or had been completed of nine petroleum-based facilities located at Angarsk (design capacity of 120,000 tonnes per annum of microbial protein), Kirishi (100,000 tonnes), Kremenchug (120,000 tonnes), Kstovo (initially 100,000 tonnes, then expanded to 200,000 tonnes), Mozyr (300,000 tonnes), Novopolotsk (120,000 tonnes), Svetloyarsk (240,000 tonnes), Syzran (120,000 tonnes) and Ufa (100,000 tonnes).[28] The facility at Mozyr was in fact constructed under the terms of a CMEA agreement signed in 1979 by the GDR, Cuba, Poland, the USSR and Czechoslovakia by which the Soviets agreed to supply some of the SCP products to the cosignatories in exchange for assistance provided during the construction.[29] By 1990, production of microbial protein in the USSR was officially reported to stand at an incredible 1,680,000 tonnes.[30]

According to the Soviet Union's own estimates, this was roughly equivalent to the addition of 8.4–11.8 million tonnes of grains to feed supplies. At a time when the Soviet Union was experiencing huge losses of agricultural produce through wastage, *Glavmikrobioprom*'s SCP output clearly represented a significant supplement to domestic protein production.

It is fascinating to learn that during the 1970s–1980s, *Glavmikrobioprom*'s microbial protein programme was being closely monitored by the CIA. Its first top secret report was compiled in April 1977 and focused on Soviet petroleum-based microbial protein production capabilities. The report estimated that "*the capacity of the completed plants is in excess of 860,000 metric tonnes, possibly one million metric tonnes annually*".[31] This was followed-up by a second secret report generated in March 1984 which identified 59 operating cellulose-based SCP plants, with one under construction, along with six operating petroleum-based SCP plants, and another under construction. One additional plant was identified which used both these feedstocks to produce microbial protein. Using imagery analysis in conjunction with reports of capacities in the Soviet open-source literature, the CIA were able to identify a total of 82 complete fermentation tanks in petroleum-based microbial protein plants in 1983. In addition, 21 fermentation tanks were under construction and were projected to be completed in 1985. The CIA used this data to estimate that the Soviets had the capacity to produce around 1.8 million metric tonnes of microbial protein in 1983 and that this figure would increase to about 2 million metric tonnes by 1985. Based on Soviet claims, the CIA analysts estimated that 2 million metric tonnes of microbial protein would be sufficient to replace as much as 10 million tonnes of grain, or about 5 per cent of the USSR's annual grain production.[32] Meanwhile, Gradon Carter of the UK's Chemical Defence Establishment, in an entirely separate study based on searches of the Soviet literature, was able to identify 86 SCP plants and, in 1981, estimated total production of microbial proteins at 1,654,265 tonnes per annum.[33]

Igor' Valerianovich Domaradskii, who was a leading *Biopreparat* scientist, has argued that *Glavmikrobioprom* was created "*in particular as a cover for connections with the Ministry of Defence and the Military Industrial Commission*".[34] However, it is clear from the analysis presented above that that this contention is simply not supported by the facts. *Glavmikrobioprom* succeeded in creating the largest microbial proteins' industry the world has ever seen which rapidly became an object of interest for Western intelligence agencies. It clearly had enormous strategic significance since, as is evidenced above, in any East-West conflict, it would have reduced the USSR's reliance on grain imports from North and South America, Australia and France. It would also put the Soviet Union in possession of a national microbial protein source that was independent of the potentially deleterious impact of the weather.[35] It is therefore apparent that *Glavmikrobioprom* at this time, as was the case in other industrial sectors in the USSR, incorporated civil programmes of global significance running contemporaneously, and sometimes, indeed, directly alongside, illicit military endeavours.

From Chemical to Biological Weapons: The Appointment of Vasilii Dmitrievich Belyaev as Head of the Microbiological Industry

It is highly significant that in 1966, the Soviet government appointed Vasilii Dmitrievich Belyaev, a former Deputy Minister of the USSR Ministry of the Chemical Industry and a figure intimately associated with the military-industrial complex, to head the newly created *Glavmikrobioprom*. Belyaev's rise to his new position was quite astonishing. He was born of very humble origins on 23 January 1916, in the village of Belyaevka, located 7 km outside Khomutovo in Orel *oblast'*, Russian Federation. In April 1932, his father, Dmitrii Nikolaevich, who was a railway worker, died, leaving behind five children. Belyaev and his family were then forced to take up residence in a "*poky little room*" in the industrial city of Kramatorsk in the Donbass. A year later, Belyaev attended a local Rabfak (*Rabochii fakul'tet*) or worker's faculty. These institutions had first been established in Soviet Russia in 1919 as a tool for social transformation. Their professed goal was to remove educational inequalities resulting from class status and to rapidly prepare peasants and workers for higher education.[36]

Later, Belyaev enrolled in the Lensovet Leningrad Technological Institute. This higher educational institution appears to have administered a special programme to train specialists in chemical weapons production. It appears likely that Belyaev was a recipient of such training at this time. It was as a student that in 1940, after the onset of the Second World War, Belyaev enrolled as a volunteer in the Red Army and participated in military campaigns on the Karelian front.[37] After graduating from the Technological Institute, Belyaev was posted to the *Zavodstroi* Chemical

Factory (Factory No. 96), in Dzerzhinsk, located on the Oka river some 35 km to the west of modern-day Nizhny Novgorod. He was initially appointed as an engineer in Shop No. 14 and went on to eventually become deputy chief engineer of Factory No. 96 and head of its Production-Technical Department. Fedorov reports that during the war, Shop No. 3 at the Zavodstroi factory was focused on the production of the chemical warfare agent, Levenstein mustard. The production areas in the factory were especially hazardous at this time because of the acute shortage of protective equipment. In 1942 alone this resulted in 1585 cases of occupational illness in the special shops and 112 cases in those focused on civil chemicals.[38]

Belyaev's close association with Soviet chemical weapons development and production was not to end here. At the end of the summer of 1948, he was summoned to a meeting with the minister of the chemical industry, Mikhail Georgievich Pervukhin. A former colleague, S.Ya. Fainshtein, from the *Zavodstroi* factory, who had been chief engineer there through to 1941, also attended the meeting. The latter was now head of the USSR Ministry of the Chemical Industry's First Main Administration. The minister offered Belyaev the post of chief engineer within Fainshtein's administration. However, Belyaev turned down the offer of this important position within the ministry and chose instead to take up an appointment as the director of Factory No. 91 in Stalingrad.[39]

Chemical Factory No. 91 was located in Beketovka to the south of Stalingrad. It had originally come on stream in 1929, and in the years through to the outbreak of the Second World War had produced mustard agent. Some 6000 metric tonnes of Levenstein mustard are reported to have been manufactured during the war.[40] Tucker notes however that Factory No. 91 was in fact a dual-purpose industrial complex with civil commercial chemicals accounting for around 65 per cent of output. According to him, "*The Soviet authorities had decided to integrate military and civilian activities at the sprawling site along the Volga in order to generate needed income and create a legitimate cover story that would mislead foreign intelligence services*".[41] Belyaev's intimate hands-on experience of running a dual-purpose facility would clearly later make him the perfect candidate for the role of head of *Glavmikrobioprom*, which would itself come to incorporate a network of dual-purpose microbiological facilities.

It is reported by Tucker that, in September 1946, the German Tabun and Sarin nerve gas plants captured by the Red Army at Dyhernfurth (now part of Poland and renamed Brzeg Dolny) were dismantled and transferred to Factory No. 91. In addition to the plants themselves, around a dozen chemists and process engineers from Dyhernfurth, who had been captured by the Red Army, were also relocated to Stalingrad. The reconstruction of the Tabun production line is reported to have proceeded smoothly, and by 1948, the agent was being stockpiled by the Red Army.[42]

Tucker further reports that after the successful production of Tabun, the Soviet Union now "*moved on to the more challenging task of manufacturing Sarin on an industrial scale*".[43] In May 1952, the CPSU Central Committee and the USSR Council of Ministers issued a secret resolution regarding the construction of a full-scale Sarin manufacturing facility at Chemical Factory No. 91. The anticipated capacity of the facility was projected to be 2000 metric tonnes per month. However, the complexities of the Sarin manufacturing process meant that full-scale production was not achieved until seven years later in 1959.[44] Belyaev's success in managing Factory No. 91 and the Sarin project was recognized just a year later in 1960 with the award of the Soviet Union's most prestigious honour, the Lenin Prize.[45]

In 1959, Belyaev was appointed as the deputy for industry of the Nizhnyi-Vol'zhsk regional economic council (*sovet narodnogo khozyaistva* or *sovnarkhoz*). As alluded to previously, *sovnarkhozes* were introduced by Nikita Khrushchev in May 1957 in what was to be a short-lived and failed attempt to combat the centralization and departmentalism of the national branch ministries.[46] Subsequently, Belyaev was promoted to the position of first deputy chairman of the *sovnarkhoz*. It is highly likely that Factory No. 91 had been transferred to the control of the local *sovnarkhoz* at this time which meant that Belyaev retained effective control of both its civil and military facilities. Another clear sign of Belyaev's burgeoning reputation in the Soviet hierarchy was the fact that in the summer of 1963, he was introduced to Nikita Sergeyevich Khrushchev, the First Secretary of the CPSU, who was undertaking a visit to Volgograd (changing its name from "Stalingrad" in November 1961).[47]

Belyaev was now to make one of the most significant moves in his career, and one which would certainly boost his prospects of securing his eventual appointment as head of *Glavmikrobioprom*. In 1963, Belyaev took up the position in Moscow as head of the Chemical Industry Department within the Central Committee of the CPSU.[48] This placed him at the very heart of Soviet policy-making and he is likely in this position to have had a significant role in oversight of the USSR's chemical weapons programme. It is no coincidence that Belyaev's two successors at

Glavmikrobioprom, Valerii Alekseevich Bykov and Rotislav S. Rychkov, had both also previously served on the staff of the Central Committee. In the early 1970s, Rychkov was a sector head within the Defence Department of the CPSU Central Committee.[49] Bykov meanwhile, had been appointed in 1979 as the head of the Microbiological Industry Section of the CPSU Central Committee's Chemical Industry Department.[50]

In 1965 Belyaev was appointed as a deputy minister of the USSR Ministry of the Chemical Industry.[51] It was in this position, that, in February 1966, he embarked upon a working visit to the Federal Republic of Germany, Holland and the UK. During the course of this trip, he was suddenly ordered to return to Moscow immediately; he later learnt that he had been appointed as head of the newly created *Glavmikrobioprom*.[52] Alexei Kosygin, the Chairman of the USSR Council of Ministers, is reported to have been a major figure of support for Belyaev's work in the *glavk*, making numerous critical interventions on Belyaev's behalf. Alongside the support from the party and the military-industrial complex, Belyaev's personal qualities were crucial in the rapid development of the new sector. Although not having a scientific background, he is reported to have been intelligent, strong-willed and a competent organizer and a good manager of personnel, able to reconcile opposing points of view. Domaradskii notes that he was intelligent and generous and showed respect to his scientists.[53]

Belyaev died on 17 July 1979 following a serious illness. It is worth examining the signatures on his official obituary which appeared in *Pravda* two days later, since these can often cast light on the pivotal connections an individual had within the Soviet system of power. The obituary is signed first by Leonid Brezhnev, the CPSU Central Committee's First Secretary, as one would expect in recognition of Belyaev's leadership of an important *glavk*. It is also signed by Aleksei Kosygin, who as we have already demonstrated, was one of Belyaev's key links within the Central Committee. The name of Mikhail Gorbachev, a rising star in the upper echelons of the CPSU, is also there. The importance of Belyaev's linkages to the chemical industry is highlighted by the signatures of Leonid Arkad'evich Kostandov, Minister of the USSR Chemical Industry and Vladimir Vladimirovich Listov, head of the CPSU Central Committee's Department of the Chemical Industry. Highlighting the crucial links between *Glavmikrobioprom* and the USSR Academy of Sciences are the signatures of Anatolii Petrovich Aleksandrov (President of the Academy), Yurii Anatol'evich Ovchinnikov and Konstantin Georgievich Skryabin. Notable among the other signatories are Rotislav Rychkov, Belyaev's successor as head of *Glavmikrobioprom*, and Vsevolod Ivanovich Ogarkov, who had headed *Biopreparat* under his tenure (see below).[54] Belyaev's role as a lead figure in the civil and military chemical and biological sectors is commemorated in Russia today by the annual award of the Vasilii Dmitrievich Belyaev Prize by the Russian Academy of Medical-Technical Sciences (*Rossiiskoi Akademii mediko-tekhnicheskikh nauk*, abbreviated to *RAMTN*).

BW R&D and Production Programmes Directly Pursued by *Glavmikrobioprom*

Domaradskii points to Belyaev's close participation in the project to manufacture Sarin nerve gas and argues that "*he always was connected with the Ministry of Defence and the Military Industrial Commission*".[55] Certainly, Belyaev's previous postings at the very heart of the Soviet chemical weapons programme made him an apposite choice for managing the Soviet network of industrial production facilities incorporating BW mobilization facilities. He was joined within the *Glavmikrobioprom* senior management team by former military personnel recruited from the USSR Ministry of Defence's Scientific-Research Technical Bureau. These included Anatolii Aleksandrovich Skladnev who was appointed head of the Technical Department and Galina Yakovlevna Shinko, who took up the position of scientific secretary of the Scientific-Technical Council.[56] The military background of *Glavmikrobioprom*'s senior leadership team was to prove useful in the management of the BW networks and programmes pursued by the *glavk*. In the first instance the dual-use plants at Berdsk, Omutninsk, Penza and Kurgan were absorbed by *Glavmikrobioprom* upon its creation in 1966. Thus, from its very inception, the new *glavk* was engaged in the management of both civil and military biological programmes of global significance.

In addition to absorbing the pre-existing network of BW production facilities, at this time *Glavmikrobioprom* also launched its own BW research programme focused on plasmids. Soviet researchers wished to use plasmids to transfer new genetic material into pathogens which would lead to an increase in virulence of classical BW agents. The new research effort was led by a leading pathogens scientist, Igor' Valerianovich Domaradskii who claimed to have been the first in the world to discover plasmids in *Yersinia pestis* (the causative organism of plague). He had left his position as director of the Rostov-on-Don Anti-Plague Institute and joined

Glavmikrobioprom in the winter of 1972. He had originally sought to establish a plasmid laboratory in the newly created *VNIIgenetika* (see above). However, for reasons unknown, this move was strongly opposed by the institute's director, Alikhanyan. As a consequence, in the autumn of 1973, Domaradskii was forced to establish his new lab in *VNIIsintezbelok* (Moscow)—an institute (see above) belonging to the *glavk* which was focused on microbial proteins production and completely lacking in any expertise in molecular biology or genetics. Belyaev pressed Domaradskii to enlarge his laboratory and he was placed in charge of an expanded Department of Extrachromosomal Inheritance. Additional staff were recruited, and eventually 50–60 personnel were employed on the plasmids programme at this facility. In addition, a bacterial pathogens plasmids lab was created by Domaradskii in Saratov and he organized a similar laboratory at the Krasnodar Branch of *VNIIsintezbelok*. Again, the latter had no prior expertise in molecular biology and up until this point had been solely focused on aspects of microbial proteins production. Belyaev also provided funds which allowed Domaradskii to contract out research to Professor Aleksandr Ivanovich Korotyaev at the Russian Soviet Federative Socialist Republic (RSFSR) Ministry of Health's Kuban State Medical Institute—named after the Red Army—in Krasnodar and Ain Heinaru at Tartu University in Estonia.[57]

After the creation of *Biopreparat*, Domaradskii's plasmid programme quickly came to the attention of its secret department. The organization appointed an observer, L.A. Melnikov, to monitor the programme. The individual was an old school friend of Domaradskii and subsequently appears to have assisted with the programme. All of Domaradskii's reports were sent to *Biopreparat* and he was forbidden to publish in any Western journals, although this prohibition was lifted after careful review with regard to publications in the Council for Mutual Economic Assistance (CMEA) countries. Despite the difficult and intrusive nature of the interaction with *Biopreparat*, Domaradskii states that his laboratory in fact was directly responsible to Belyaev who approved all of his plans and reports. The head of *Glavmikrobioprom* also ensured that Domaradskii was protected from intrusive questions about his work from the *VNIIsintezbelok* hierarchy.[58]

Initially Domaradskii's research was focused on the study of plasmids in the bacteria *Escherichia coli* and *Pseudomonas*. In 1976, Belyaev personally applied to the Military Industrial Commission (*VPK*) and the Central Committee of the Communist Party for permission for Domaradskii's laboratory to begin research aimed at the transfer of foreign genes to plague bacteria. The intention was to conduct research using the EV avirulent strain. Permission was required because the Soviet authorities had previously instigated a ban on plague research within the environs of Moscow following an accidental release of the microorganism in the city in December 1939. According to Domaradskii, this project at *VNIIsintezbelok* was approved by none other than the head of the KGB, Yurii Andropov.[59] The fact that Andropov was willing to authorize the plague research in contravention of a long-standing official prohibition on such activity in Moscow is a vivid illustration of the importance attached to the project by the Soviet authorities.

Domaradskii reports that some "cleared" members of *VNIIsintezbelok* refused to participate in the proposed programme of work on the EV plague strain. However, despite this the project proceeded and he was successful in transferring a cloned haemolytic plasmid from *Escherichia coli* into colonies of EV. He notes that "*our experiment ...proved that alien genetic material could be introduced into the plague bacillus, something that had never been demonstrated before*".[60] This technology he argues was the basis of a military bioweapons programme, permitting the creation of strains with new properties. Domaradskii notes that the most significant achievement of his *VNIIsintezbelok* lab occurred in 1977 when he and his team succeeded in demonstrating the existence of plasmids in *Yersinia pestis* and the central role they played in plague virulence. He claims that it took another three years before Western scientists were to make this same discovery.[61]

During the course of its existence, there do appear to have been some severe security lapses associated with the plasmid project. In February 1977, for example, V. Kotov, a member of *Glavmikrobioprom*'s Foreign Relations Department, wrote to Reg Bottomley of Peruvum Limited (UK) regarding a joint seminar and exhibition (see Fig. 3.3).[62] It was envisaged that the event would take place from 30 May to 3 June of that year at the All-Union Scientific-Research Institute of the Biosynthesis of Protein Substances (*VNIIsintezbelok*) in Moscow.[63] Astonishingly, Kotov proposed that one of the two topics for the seminar should focus on "*Genetics of plasmids (profs. Datta and Hedges from Medical School of London, prof. Willets from University of Edinburgh, prof. Meynell from University of London)*".[64] Presumably, Domaradskii had pressured *Glavmikrobioprom* to present this request to the UK. However, it does appear incredible that *VNIIsintezbelok* should be publicly linked in this way, albeit not directly, to the plasmids programme. Is it the case that this

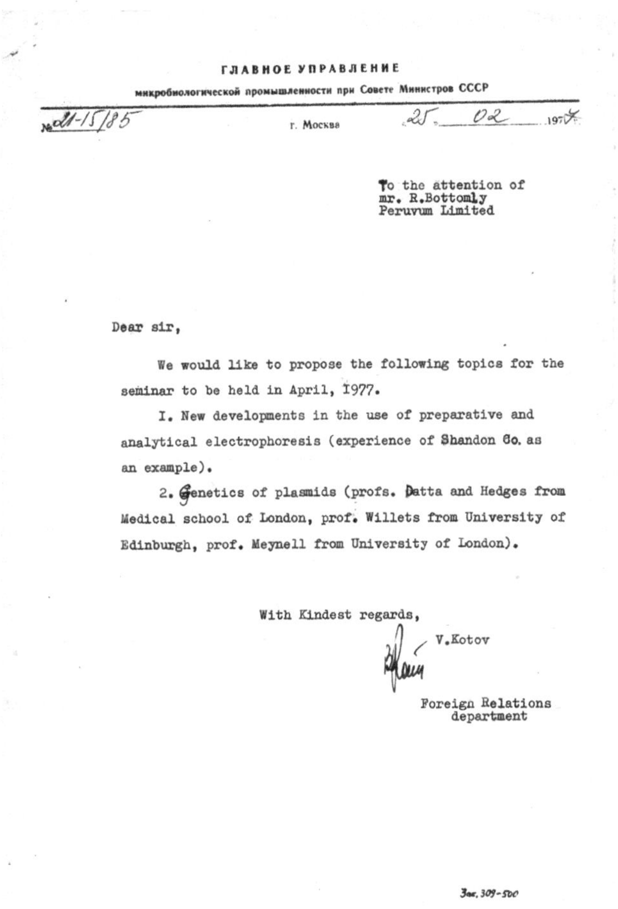

ГЛАВНОЕ УПРАВЛЕНИЕ
микробиологической промышленности при Совете Министров СССР

№ 21-15/85 г. Москва 25. 02 1977

To the attention of
mr. R.Bottomly
Peruvum Limited

Dear sir,

We would like to propose the following topics for the seminar to be held in April, 1977.

I. New developments in the use of preparative and analytical electrophoresis (experience of Shandon Co. as an example).

2. Genetics of plasmids (profs. Datta and Hedges from Medical school of London, prof. Willets from University of Edinburgh, prof. Meynell from University of London).

With Kindest regards,
V.Kotov

Foreign Relations
department

Зак. 309-500

Fig. 3.3 Letter regarding plasmids research from *Glavmikrobioprom*'s Foreign Relations' Department to Reg Bottomley, Peruvum Ltd. (UK), 25 February 1977 (©A. Rimmington)

was viewed as an unprecedented opportunity to procure knowledge of the West's scientific capabilities with regards to plasmids and concerns with regard to exposing Domaradskii's secret laboratory were put to one side? It is not known if Western intelligence were ever alerted to this expression of *Glavmikrobioprom*'s interest in the UK's research on the genetics of plasmids, but clearly this would have been a strong indicator of the direction of travel of Soviet bioscience R&D at this time.

Domaradskii was subsequently able to gain official support for an All-Union Plasmid Programme from the Interagency Scientific-Technical Council for Problems of Physical-Chemical Biology and Biotechnology, which was subordinate to the State Committee for Science and Technology and the Presidium of the USSR Academy of Sciences. *Biopreparat* leant additional support with the funding of conferences and publications devoted to this subject. Beginning in 1978, he was able to organize meetings under this programme at cities across the Soviet Union. From 25 to 27 September 1980, for example, the fifth All-Union plasmids meeting was held in Krasnodar on *Plasmid Structure and Function*.[65] The ninth meeting of the Plasmid Programme on *Hybrid Plasmids and Expression of Plasmid Genes* was organized from 18 to 21 November 1984 by Domaradskii at the USSR Academy of Sciences' Institute of Biochemistry and Physiology of Microorganisms (*IBFM*) in Pushchino, outside Moscow.[66] Other locations for Plasmid Programme conferences were Kiev, Tartu and Saratov. The situation vis- à-vis the plasmid conferences became more difficult after Belyaev's death in July 1979. They were suspended after 1987, but Domaradskii managed to secure the support of Adelina-Viktoriya Genrikhovna Skavronskaya at the USSR Academy of Medical Sciences' N.F. Gamaleya Institute of Microbiology and Epidemiology (Moscow). This allowed a final conference to be held in Nalchik, the capital city of the Kabardino-Balkaria Republic, in 1990.

One of Domaradskii's meetings in Tallinn in October 1982 on *Metabolic Plasmids* was attended by 22 foreign scientists from nine countries, together with 110–120 Soviet participants, although it is uncertain how many of the latter were scientists. The event is reported to have been sponsored by *Biopreparat*. The international scientists included Raymond Zilinskas from the US, who was invited to attend by *Glavmikrobioprom* and would later go on to co-author a definitive account of the Soviet BW programme.[67] All of the international participants, even including those from the USSR's CMEA allies, were accommodated separately from the Soviet scientists. Zilinskas notes, "*While conference's attendees met at some social events, these were strained encounters because most of the Soviets did not speak English and they presumably were being monitored by the KGB*".[68] Presumably, a careful watch was kept on Zilinskas and his colleagues throughout his participation in this event. As a result of his interaction with the foreign scientists, Domaradskii subsequently received an invitation to attend as a co-chair and speaker at an international symposium on molecular genetics in the US. However, he was refused permission by the Soviet authorities and "*Once again, I had to 'report sick' at the last moment*".[69] Zilinskas argues that the international nature of the event in Tallinn was an attempt by Domaradskii to confirm to Western observers, *Biopreparat*'s credentials as a civil institution.

The Creation of Microbiological Facilities at Stepnogorsk

In 1970, Vasilii Belyaev played a crucial role in initiating the creation of new microbiological facilities at a site some 14 km north of Stepnogorsk in Kazakhstan. The project is of immense significance because, later, it involved the construction of a major military component, the so-called Scientific Experimental-Industrial Base or *SNOPB*, which was eventually to achieve production capacity of 300 metric tonnes of weaponized *B. anthracis* spores per ten-month production cycle.[70] The creation in Stepnogorsk of the new Chemical Factory *Progress* (*Khimicheskii zavod "Progress"*)—the most powerful in the Soviet Union—was originally envisaged in Decree No. 664 "*Concerning measures for the accelerated development of the microbiological industry*" issued on 8 August 1970 by the Central Committee of the CPSU and the USSR Council of Ministers.[71] In accordance with the decree, Belyaev issued the following order on 30 December 1970: "*With the aim of supplying agriculture with microbial pesticides...during 1971 to 1975 Glavmikrobioprom are to carry out the construction of the Chemical Factory Progress in the town of Stepnogorsk, Tselinograd oblast*"'.[72]

A number of factors influenced the Soviet authorities in their decision to locate a major new BW facility in north-east Kazakhstan including its extreme remoteness but also the availability of powerful construction bases and local utilities developed by the Soviet defence complex. The town of Stepnogorsk itself was located around 143 kilometres to the North of Tselinograd (modern-day Nur-Sultan), close to the border of the adjacent Kokchetav *oblast'*. The town originated as a result of the opening of a new mining and chemical combine for the extraction of uranium by the USSR Ministry of Medium Machine Building (*Minsredmash*) in 1956. The Minister of *Minsredmash*, Efim Pavlovich Slavskii, personally selected the site for the construction of Stepnogorsk. *Minsredmash* was part of the

Soviet defence complex and was responsible for production of strategic and intermediate-range ballistic missiles.[73] Stepnogorsk was formerly a "closed" town and prior to the collapse of the USSR, it was referred to under several names including "Makinsk-2", "Tselinograd-25" and "Aksu". In June 1964, an unpublished decree of the Presidium of the Supreme Soviet of the Kazakh SSR conferred the status of a town on Aksu and it was given the name Stepnogorsk. Among the other production plants which were subsequently built at Stepnogorsk were a Preserved Milk Products Combine, a Bearings Factory and a Ceramics Factory.

The domination of Stepnogorsk by *Minsredmash* is of particular interest since the origins of the microbiological complex in Stepnogorsk do appear to lie in the excellent personal relations which existed between Belyaev, the head of the Main Administration of the Microbiological Industry (*Glavmikrobioprom*), and Efim Pavlovich Slavskii of *Minsredmash*. Moreover, their respective deputies responsible for construction, Robert Sergeevich Zurabov and Veniamin Evseevich Lyubomirskii, were close friends. It was Slavskii who made available *Minsredmash*'s excess construction capacity to *Glavmikrobioprom* for the erection of its first factory in Stepnogorsk.[74] Appendix No. 8 to the 1970 Central Committee decree concerning the accelerated development of the microbiological industry confirms that the construction of *Progress* was the responsibility of *Minsredmash*.[75] This was one of the only plants in the entire *Glavmikrobioprom* production network that was constructed by the nuclear weapons' ministry and is perhaps the crucial indicator of the special status it was subsequently to hold in the Soviet offensive BW programme.

Progress's first director, Yurii Nikolaevich Timon'kin, was appointed under the terms of *Glavmikrobioprom* Order No. 17 issued on 14 December 1970. He arrived at the Stepnogorsk site in January 1971. *Glavmikrobioprom* initially allocated between 170–180 million roubles (1970 prices) for the construction of *Progress* which was intended to be one of the largest and technologically sophisticated microbiological facilities in the former Soviet Union. By 31 December 1971, the new plant only had some 20 employees. Timon'kin then recruited just under 100 microbiologists, chemists and other staff from his previous facility, the dual-use plant in Berdsk—clearly strong bonds of loyalty had been formed during his period as director there.[76] Staff were also recruited from the dual-use facilities in Penza and Kurgan and other civil microbiological plants. The availability of apartments in Stepnogorsk was a major reason why large numbers of qualified personnel relocated to the remote outpost. As an important defence-related centre, its stores were also well-stocked and had priority access to foodstuffs and other products. A tranche of graduates from the Leningrad Chemical-Pharmaceutical Institute were also recruited after a visit by Timon'kin. Staff numbers rose rapidly from 231 people in 1973, to 1502 people in 1974 and to 2035 people in 1975.[77] Eventually, the number of staff in Stepnogorsk reached around 4000 personnel. Many of the younger members of staff were sent on internships to the plants in Berdsk and Kurgan, where they studied fermentation processes and other operations in situ. Convict labour sourced from a penal colony located 3 km away, was used in the construction of the Stepnogorsk facility. Timon'kin built a small hotel at the site for the use of visiting specialists and others and a sports complex. In 1971, construction began of Units Nos. 1, 2 and 6 for the production of 10,000 tonnes per annum of Enterobakterin, a microbial pesticide based on *Bacillus thuringiensis*. The first industrial batch of Enterobakterin was produced in Unit No. 1 on 23 January 1975. Construction of Units Nos. 3, 5, 7 and 8 for the production of 200 tonnes per annum of antibiotics for use in animal feed began in 1972. However, lack of consumer awareness and demand for the latter products later led the management to reequip the antibiotics units for the output of the essential amino acid lysine—a key supplement in animal feed.[78] Eventually, *Progress* incorporated 122 63,000-litre fermenters, manufactured at a plant in Karl-Marx-Stadt (present-day Chemnitz) in the GDR.[79] Timon'kin, who died in 2007, was eventually replaced in 1983 by Eduard Ivanovich Perov, a former employee at *Glavmikrobioprom*'s Berdsk Chemical Factory. The latter had been recruited by Timon'kin and transferred to Stepnogorsk in 1971. He had been employed in various senior posts at *Progress* including chief mechanic and power engineer, and then deputy director for experimental production.[80]

According to Leitenberg and Zilinskas, from its very inception, the Stepnogorsk site incorporated both a military and civilian component. The civil plant, *Progress*, occupying 170 hectares, was largely completed by the time construction of the military component, *SNOPB*, began on the immediately adjacent land, occupying 100 hectares. The two sites were separated by a fence but shared key infrastructure such as a power plant and a steam plant. *Progress* possessed its own Design and Construction Department (*Proektnyi-konstruktorskii otdel* or *PKO*) which was created on 6 February 1973. Its first director was A.V. Kupreev. *PKO* later played a major role in the construction of new production facilities at the neighbouring *SNOPB* complex (see below).

The Establishment of the Interdepartmental Scientific-Technical Council (*MNTS*) for Molecular Biology and Genetics Under *Glavmikrobioprom*

In 1971, *Glavmikrobioprom* established a new top-secret nerve centre for the management of the renewed Soviet BW programme, the Interdepartmental Scientific-Technical Council for Molecular Biology and Genetics (*Mezhdomstvennyi nauchno-tekhnicheskii sovet po molekulyarnoi biologii i genetike*, abbreviated for convenience in the text as *MNTS*)—see Appendix B. The Council which was initially headed by a virologist, Academician Viktor Mikhailovich Zhdanov, included representatives from a range of military and civil agencies participating in a new, expanded, BW programme. The USSR Ministry of Defence's Fifteenth Administration was represented by its deputy head at this time, Major General Vladimir Andreevich Lebedinskii. From the USSR Academy of Sciences, there were three of the Soviet Union's most renowned and talented bioscience researchers, Yurii Anatol'evich Ovchinnikov, Georgii Konstantinovich Skryabin and Aleksandr Aleksandrovich Baev. Other departments with representation included the USSR Ministry of Agriculture and the USSR Ministry of Health, whose deputy minister and head of its Third Main Directorate, Lieutenant-General (Reserves) Avetik Ignat'evich Burnazyan, sat on the council. Belyaev also ensured that *Glavmikrobioprom* was well-represented on the *MNTS*. Besides himself, both Anatolii Aleksandrovich Skladnev (head of *Glavmikrobioprom*'s Technical Administration) and Sos Isaakovich Alikhanyan (*VNIIgenetika*—see above) were also members. High-level representatives from the Central Committee of the Communist Party, the *VPK* and the KGB always attended meetings of the Council. In addition, Domaradskii (see immediately below) served under Zhdanov as the deputy chairman of the *MNTS*.

The sessions of the *MNTS* were held in a specially constructed soundproof auditorium which had been scanned prior to use by the security services. Military figures who attended the meetings were dressed in their civilian clothes. The members of the council received what for Soviet scientists was an extraordinary privilege—an annual hard currency allocation of US$500 for the purchase of subscriptions to scientific journals. The Council was responsible for the formulation of new scientific programmes, the coordination of the many different agencies engaged in the BW effort and the allocation of major sums of hard currency. Of the US$10 million per annum of hard currency available in the 1970s, the lion's share (40–45 per cent) went to the newly formed *Biopreparat* (see below) and around 30 per cent to the USSR Academy of Sciences, with the Ministries of Defence, Health and Agriculture each receiving around US$1–1.5 million for their BW projects from this source. However, key decisions with regard to the BW programme could only be implemented with the approval of the Central Committee and the Military-Industrial Commission (*VPK*).[81]

The *MNTS* had its own Administrative (or Special) Department which operated under the cover designation of P.O. Box A-3092 (*p/ya A-3092*)—a secret within a secret as it were. It was based initially in a small room at *Glavmikrobioprom*'s headquarters at ulitsa Lesteva, 18. From 1973 through to 1982, it was none other than Igor' Valerianovich Domaradskii (see above), who served as the head of the department. He had been recruited for his new role by the deputy health minister Burnazyan (see above) who recognized his capabilities during their joint work fighting a cholera epidemic in Nukus, Uzbekistan.[82] Domaradskii's appointment to this key position is rather unusual since he had a civil rather than a military background. Burnazyan clearly must have recognized his outstanding scientific qualities and overridden any concerns regarding security which may have arisen. Domaradskii's deputy within the Special Department was Colonel Lev Aleksandrovich Klyucherev. Described as an intelligent and decent man, he had served at the front as a medical assistant during the Second World War. Later, in 1953, he defended his Candidate's dissertation at the S.M. Kirov Military-Medical Academy (Leningrad) and was subsequently transferred to the military's Scientific-Research Sanitary Institute (Zagorsk) where he was engaged in work on rickettsia and viruses. The Special Department also employed a number of people who served as assistants to members of the *MNTS* and prepared documents and analyses for them.[83]

Domaradskii notes that the Special Department was focused on preparing documentation for the *MNTS*. This included the correction of plans for *Biopreparat*'s R&D and production, estimates of the results of scientific and technical work and so on. All of this work had to be considered by the council and approved by the Higher Certification Committee. Plans had also to be drawn up by *p/ya A-3092* for the implementation of a special order, issued annually by the USSR Ministry of Defence, regarding fundamental and applied tasks. All decisions taken by the *MNTS* required the unanimous agreement of all its members and Domaradskii, who acknowledges that he lacked diplomatic skills, often sent his deputies in his stead to attempt to reconcile the conflicting parties. Draft letters were also prepared for the chair of the *MNTS* to approve and send to the government and the CPSU.[84] Clearly, in his position as head of the Special Department, Domaradskii occupied a vitally important position at the very heart of the Soviet BW programme—a fact that makes his published memoirs an important reference point for historians attempting to document the Soviet BW programme.

In 1974, Domaradskii, in conjunction with Zhdanov, drafted an ambitious plan for the development of a new generation of biological weapons, which became known as the "Five Principal Directions". The plan was subsequently approved in 1975 by the entire *MNTS* and all the requisite political and military authorities. It then later became known as the *Bonfire* (*Koster*) programme, which itself formed part of the overall Soviet BW programme, given the codename *Problem Ferment*. Domaradskii is somewhat vague as to the precise objectives of *Bonfire* but noted in his memoirs: "*We planned to introduce new properties into disease organisms, such as antibiotic resistance, altered antigen structure and enhanced stability in the aerosol form, making delivery of the agent easier and more effective.... We also reasoned that the genes for toxins or other factors could be introduced into cells of bacteria or the DNA of viruses and thereby yield strains with wholly new and unexpected pathogenic properties*".[85] Some of the Soviet accomplishments with regard to the weaponization of pathogens, which were encompassed by the *Bonfire* programme, are discussed below.

According to Domaradskii, Belyaev always attended sessions of the *MNTS* with Zhdanov usually yielding the chair to him in deference to his position as head of *Glavmikrobioprom*. By 1975, Belyaev had come to recognize the increasingly important role the council was playing in the management and funding of the Soviet BW programme and, in particular, the newly created *Biopreparat*. He therefore implemented a major reorganization of the council. Zhdanov was dismissed as chairman and Belyaev had himself appointed in his place. He was to serve in this position until his death in the late autumn of 1979. Domaradskii was removed as deputy chairman but crucially, as a loyal and trusted confidant of Belyaev, was retained as the council's scientific secretary and retained his position as head of *MNTS*' Special Department. In addition, Colonel-General Efim

Ivanovich Smirnov (the new head of the USSR Ministry of Defence's Fifteenth Administration) and Major General (Reserves) Vsevolod Ivanovich Ogarkov (head of *Biopreparat*—see later) were appointed members.[86]

In 1979, after the death of Belyaev, Rotislav Rychkov, a communist party apparatchik, was appointed head of *Glavmikrobioprom*. He served concomitantly as chairman of the *MNTS*. Immediately, there was a major clash of personalities between Rychkov and Domaradskii. This was to lead to the latter being removed both from membership of the council and, in February 1982, as head of its Special Department. Major General (Reserves) Dmitrii Vladimirovich Vinogradov-Volzhinskii subsequently replaced Domaradskii as head of the Special Department.[87]

Notes

1. Postanovlenie Soveta Ministrov SSSR, 18 fevral' 1966g., O razvitii mikrobiologicheskoi promyshlennosti i ob organizatsii upravleniya etoi promyshlennost'yu, in *Resheniya partii i pravitel'stva po khozyaistvennym voprosom, Vol. 6, 1966—iyun' 1968 goda*, Politicheskoi literatury, Moscow, 1968.
2. Kostandova, L.A., Zhavoronkova, N.M. (eds.), *Razvitie otdel'nykh otraslei khimicheskoi promyshlennosti*, Vol. 2, Nauka, Moscow, 1984, p. 347.
3. *Directory of Soviet Officials, Volume 1: National Organizations, A Reference Aid*, National Foreign Assessment Centre, CR 78–14,025, September 1978, p. 295.
4. Letter to L.I. Brezhnev, From V. Belyaev—Head of the Main Administration, 18 February 1970, Decrees of the Central Committee of the CPSU and Council of Ministers etc., *RGAE (Russian State Archive of the Economy)*, Fond 178, Opis 1, Delo 953, 1970, pp. 2–5.
5. A Brief Historical Abstract, Homepage of USSR Academy of Sciences' Institute of Molecular Genetics, http://www.img.ras.ru/hist_e2.htm, Accessed on 15 March 2004.
6. Istoriya instituta, http://www.genetika.ru/about/hoistory/, Accessed on 3 August 2020; Rimmington, *Who's Who*, pp. 29–30.
7. Domaradskij, Orent, *Biowarrior*, p. 164.
8. Domaradskij, Orent, *Biowarrior*, pp. 145, 164–165.
9. Sagdeev, R., How the Soviets Got the Bomb, *Popular Science*, August 1994, p. 30.
10. Peredache Glavnomu upravleniyu mikrobiologicheskoi promyshlennosti tekh zdanii dlya razmeshcheniya Vsesoyuzno nauchno-issledovatel'skogo instituta mikrobiologicheskikh sredstv zashchity rastenii i bakterial'nykh

preparatov, Sovet Ministrov SSSR, No. Pr-160, 17 May 1974, *RGAE*, Fond 178, Opis' 1, Delo 2024, pp. 1–3.
11. Introduction to the *Glavmikrobioprom* archive, *RGAE*, Fond 178, Opis 1, undated document.
12. Sandakhchiev, L.S., *Innovation Development Programme of the Association "VECTOR" for 1992–1995 and till (sic) 2005*, State-Owned Concern "BIOPREPARAT" SCIENTIFIC-PRODUCTION ASSOCIATION "VECTOR", Novosibirsk, 1991, p. 16.
13. Filial FBUN GNTs VB "Vektor" Institut meditsinskoi biotekhnologii, http://www.vector.nsc.ru/filial-/, Accessed on 12 August 2020.
14. Domaradskij, Orent, *Biowarrior*, p. 144.
15. Humphrey, A.E., Single cell protein: A case study in the utilization of technology in the Soviet system, in Thomas, J.R., Kruse-Vaucienne, U.M. (Eds.), *Soviet Science and Technology: Domestic and Foreign Perspectives*, George Washington University, Washington, D.C., 1977.
16. Kozlov, A.I. Epshtein, Ya.V., Gidroliznaya promyshlennost' k 60-letiyu Velikhogo Oktyabrya, *Gidrolizпoe proizvodstvo, sbornik referativnykh materialov*, No. 11(100), 1977, p. 3.
17. Ibid. and Pokrovskii, A.A., Perspektivy ispol'zovaniya belkov odnokletochnykh organizmov, p. 13 in Pokrovskii, A.A. (Ed.), *Mediko-biologicheskie issledovaniya uglevodorodnykh drozhzhei*, Moscow 1972.
18. Kozlov, A.I. Epshtein, Ya.V., Gidroliznaya promyshlennost' k 60-letiyu Velikhogo Oktyabrya, p. 3.
19. Andreev, A.A., Bryzgalov, L.I., *Proizvodstvo kormovykh drozhzhei*, Lesnaya promyshlennost', Moscow, 1986, p. 6.
20. *USSR: Single-Cell Protein Industry (U): A Research Paper*, Central Intelligence Agency, Directorate of Intelligence, IA 84-10,027, March 1984, p. 2, https://www.cia.gov/library/readingroom/docs/CIA-RDP91T01115R000100080003-4.pdf, Accessed on 27 July 2020.
21. Humphrey, A.E., Single cell protein: A case study in the utilization of technology in the Soviet system, p. 343.
22. Rimmington, A., with Greenshields, R., *Technology and Transition: A Survey of Biotechnology in Russia, Ukraine and the Baltic States*, Quorum Books, Westport, Connecticut, 1992, p. 9.
23. Neiman, B.Ya., *Industriya mikrobov*, Znanie, Moscow 1983.
24. Katrush, R.V., Mirzayanova, E.G., Industriya belka, p. 8 in *Zhurnal' vsesoyuznogo khimicheskogo obshchestva im. D.I. Mendeleeva*, Vol. 27, No. 6, 1982.
25. Rimmington, A., Single Cell Protein: The Soviet Revolution? *New Scientist*, Vol. 106, No. 1462, 27 June 1985, p. 12.
26. Rimmington with Greenshields, R., *Technology and Transition*, p. 9.

27. Kostandova, L.A., Zhavoronkova, N.M. (eds.), *Razvitie otdel'nykh otraslei khimicheskoi promyshlennosti*, p. 349.
28. Rimmington, A., *Biotechnology in the USSR*, Thesis submitted for the Degree of Doctor of Philosophy, Centre for Russian and East European Studies, University of Birmingham, 1988, pp. 85–87.
29. Skomorokhin, Y., Sostavnoe zveno agropromyshlennogo kompleksa, *Ekonomicheskoe Sotrudnichestvo Stran—Chlenov SEV*, No. 1, 1984, p. 38.
30. *Narodnoe khozyaistvo SSSR v 1990 g.*, Finansy i statistika, Moscow, 1991, p. 412.
31. *The Soviet Hydrocarbon-Based Single Cell Protein Programme*, SI 77-10014K, Central Intelligence Agency, April 1977, p. i.
32. *USSR: Single-Cell Protein Industry (U): A Research Paper*, pp. 1–8.
33. Carter, G.B., Is Biotechnology Feeding the Russians?, *New Scientist*, Vol. 90, No. 1250, 23 April 1981, p. 218.
34. Domaradskij, Orent, *Biowarrior*, pp. 117–118.
35. Carter, G.B., Is Biotechnology Feeding the Russians?, p. 218.
36. Miethe, I., *The Worker's Faculty: The Globalisation of Soviet Education*, https://www.scientia.global/wp-content/uploads/Ingrid_Miethe/Ingrid_Miethe.pdf, Accessed on 29 July 2020.
37. Vasilii Dmitrievich Belyaev, *Pravda*, 19 July 1979.
38. Fedorov, L., *Chemical Weapons in Russia: History, Ecology, Politics*, Moscow, 1994, https://fas.org/nuke/guide/russia/cbw/jptac008_l94001.htm, Accessed on 29 July 2020.
39. *Plamennoe serdtse*, History of the life of Vasilii Dmitrievich Belyaev, unpublished manuscript, Moscow, 2006.
40. Fedorov, L., *Chemical Weapons in Russia*, loc cit.
41. Tucker, J., *War of Nerves: Chemical Warfare from World War I to Al-Qaeda*, Pantheon Books, New York, 2006, pp. 106–107.
42. Ibid.
43. Ibid.
44. Fedorov, L., *Chemical Weapons in Russia*, loc cit.
45. *Plamennoe serdtse*, loc cit.
46. Hanson, P., *The Rise and Fall of the Soviet Economy: An Economic History of the USSR from 1945*, pp. 58–60.
47. N.S. Khrushchev na nizhnei volge, *Pravda*, summer of 1963, photocopy kindly provided by scientist close to Belyaev, precise date of publication unknown.
48. Vasilii Dmitrievich Belyaev, *Pravda*, 19 July 1979; *Plamennoe serdtse*, loc cit.
49. Rychkov, R.S., Aktual'nye problemy razvitiya mikrobiologicheskoi promyshlennosti, *Zhurnal Vsesoyuznogo khimicheskogo obshchestva im. D.I. Mendeleeva*, Vol. 27, No. 6, 1982, pp. 13–17.

50. Rimmington, *Who's Who*, pp. 56–57.
51. K 100-letiyu so dnya rozhdeniya V.D. Belyaeva—organizatora mikrobiologicheskoi promyshlennosti SSSR, *Vestnik biotekhnologii i fiziko-khimicheskoi biologii imeni Yu.A. Ovchinnikova*, Vol. 12, No. 3, 2016, p. 54.
52. *Plamennoe serdtse*, loc cit.
53. Domaradskij, Orent, *Biowarrior*, p. 194.
54. Vasilii Dmitrievich Belyaev, *Pravda*, 19 July 1979.
55. Domaradskij, Orent, *Biowarrior*, pp. 117–118.
56. Mityushev, *Zapiski obyknovennogo cheloveka*, p. 416.
57. Domaradskij, Orent, *Biowarrior*, p. 167.
58. Ibid., p. 166.
59. Ibid., pp. 173–174.
60. Ibid., p. 175.
61. Ibid., pp. 175–182.
62. Letter No. 21-15/85 from V. Kotov, Foreign Relations Department of *Glavmikrobioprom* to Mr Reg Bottomly, Peruvum Limited, 25 February 1977. This letter forms part of Mr Reg Bottomly's personal archive of Peruvum business documentation which was later presented to Dr Rimmington. Peruvum Ltd. acted as consultants for a number of UK-based biotechnology equipment suppliers and during the 1970–1980s, undertook a significant number of business visits to the Soviet Union. Detailed reports were compiled for UK clients and these reports, together with some correspondence, form the bulk of the archive.
63. Letter No. 21-15/37 from M. Zhed', Head of the Foreign Relations Department of *Glavmikrobioprom* to Mr Reg Bottomly, Peruvum Limited, 27 January 1977.
64. Letter No. 21-15/85 from V. Kotov, 25 February 1977.
65. *Struktura i funktsii plasmid, Tezisy dokladov, U Vsesoyuznoe soveshchanie po programme "Plasmide"*, Krasnodar, 25–27 September 1980, Moscow, ONTITEImikrobioprom, Moscow, 1980.
66. *Gibridnye plazmidy i ekspressiya plamidnykh genov, Tezisy dokladov, IX Vsesoyuznyi soveshchanie po programme "Plasmida"*, USSR Academy of Sciences' Institute of Biochemistry and Physiology of Microorganisms, 18–21 November 1984, VNIISENTI, Moscow, 1984.
67. Zilinskas, R., Biotechnology in the USSR, Part 1, *Bio/Technology*, July 1984, pp. 610–615.
68. Leitenberg, Zilinskas, with Kuhn, *The Soviet Biological Weapons Programme*, p. 159.
69. Domaradskij, Orent, *Biowarrior*, p. 169.
70. Leitenberg, Zilinskas, with Kuhn, *The Soviet Biological Weapons Programme*, p. 275.

71. O merakh po uskoreniyu razvitiya mikrobiologicheskoi promyshlennosti, Decree No. 664 of the Central Committee of the CPSU and the USSR Council of Ministers, 8 August 1970, *RGAE*, Fond 178, Opis' 1, Delo 953, pp. 1–73.
72. Perov, E.I., Trudnye shagi "Progressa" pp. 107–125 in *Stepnogorsk. Zdes' propisany nashi serdtsa. Stranitsy istorii goroda Stepnogorska Akmolinskoi oblasti*, Atamura, Almaty, 1994.
73. Malleret, T., *Conversion of the defence industry in the former Soviet Union*, Institute for East-West Security Studies, Occasional Paper No. 23, Westview Press, New York, 1992; Cooper, J.M., Civilian and military engineering, in Berry, M.J., *Science & Technology in the USSR*, Longman, Harlow, 1988, pp. 271–292.
74. Perov, Trudnye shagi "Progressa", pp. 107–125.
75. Perechen' predpriyatii Glavmikrobioproma, stroitel'stvo kotorykh vozlagaetsya na Mintyazhstroi SSSR, Minpromstroi SSSR, Minstroi SSSR, Minsel'stroi SSSR, Minenergo SSSR i Minsredmash, Prilozhenie k postanovleniyu TsK KPSS i Soveta Ministrov SSSR ot 1970, *RGAE*, Fond 178, Opis 1, Delo 953, pp. 48–49.
76. Stepnogorsk, 1970–1984, Perov, Botov i Podkopaeva, http://press.ucoz.kz/load/3-1-0-4, Accessed on 9 October 2020.
77. Timon'kin, Yu.N., *Ya tvoeyu sud'boyu zhivu.*
78. A major contribution to this work was made by the Chief Engineer Vladimir Nikolaevich Davydov who had a detailed knowledge of lysine production from his previous work in Belarus. Perov, Trudnye shagi "Progressa", pp. 107–125.
79. Timon'kin, Yu.N., *Ya tvoeyu sud'boyu zhivu.*
80. Korotko ob avtorakh, *Stepnogorsk. Zdes' propisany nashi serdtsa.*, pp. 201–202.
81. Oleg Ignat'ev is reported to have been the head of a BW department within the *VPK* during the 1970s. Domaradskii, I., Istoriya odnoi avantyury, *Znanie-Sila*, No. 11(833), 1996, pp. 67–68.
82. Domaradskij, Orent, *Biowarrior*, p. 119.
83. Ibid., pp. 146–148.
84. Ibid., pp. 145 & 154–155.
85. Ibid., p. 178.
86. Domaradskii, I., Istoriya odnoi avantyury, pp. 59–65.
87. Ibid., p. 301.

CHAPTER 4

Genesis: The Creation of *Biopreparat*

By 1970 there were already three main departments for the management of Soviet microbiological facilities in existence within *Glavmikrobioprom*. These were the Administration of the Hydrolysis Industry (*Upravleniya gidroliznoi promyshlennosti*) which focused on the production of microbial proteins from cellulose. In addition, there was the Administration of the Protein Substances and Enzymes Industry (*Upravleniya promyshlennosti belkovykh veshchestv i fermentov*) which was primarily focused on producing microbial proteins from petroleum. The third unit, which had been created by Belyaev, was the Administration of the Bacterial Preparations Industry (*Upravleniya promyshlennosti bakterial'nykh preparatov*). Subsequently, the latter became heavily focused on the production of amino acids, especially methionine—a crucial feed additive—and under the terms of *Glavmikrobioprom*'s Order No. 169 on 3 June 1974 was assigned factories based in Livany, Obol', Charentsavan and Shebekino.[1]

On 24 April 1974, *Glavmikrobioprom* issued Order 131 DSP which officially created a major new department within *Glavmikrobioprom*, the now notorious All-Union Science Production Association *Biopreparat* (*Vsesoyuznoe nauchno-proizvodstvennoe ob'edinenie "Biopreparat"* or *VNPO "Biopreparat"*).[2] The Association was also referred to in coded cables as P.O. Box A-1063. The science production associations (or *NPOs*) in the Soviet Union at this time were intended to be an "*organizational fusion of research, design and experimental production facilities and one or*

A. Rimmington, *The Soviet Union's Invisible Weapons of Mass Destruction*, https://doi.org/10.1007/978-3-030-82882-0_4

more enterprises".[3] The new department was initially based at *Glavmikrobioprom*'s headquarters at ulitsa Lesteva, 18. However, it was subsequently transferred to a building formerly belonging to the Academy of Chemical Defence located in Moscow's Lefortovo district at Samokatnaya ulitsa, 4a.[4] It is described as an elegant building with tall windows and a green roof, and it is reported to have once been the home of the Russian vodka merchant Petr Arsen'evich Smirnov.[5] It is located immediately adjacent to the *Kristall* vodka distillery and was ideally suited for a secret organization such as *Biopreparat* since it was located on a hidden driveway. Domaradskii, whose Special Department (see above) was also transferred to this location, notes that the "***building was enclosed in a solid fence and surrounded by a park; the only decent approach was from the direction of the Yauza, a tributary of the Moscow River, but it was hard to pick out***".[6] Another, *Biopreparat* employee adds that for those wishing to visit, having the postal address alone, didn't help locate the building and it was better to have a map.

The military focus of *Biopreparat* was brought into sharp focus with the appointment of Major General (Reserves) Vsevolod Ivanovich Ogarkov as director—concomitantly also taking up the position of deputy director at *Glavmikrobioprom*. Ogarkov, who was born on 2 February 1926 in the Russian city of Smolensk, was closely associated with the Soviet Army's BW programme. His initial training took place in the S.M. Kirov Military Medical Academy in Leningrad. His father, Colonel Ivan Fedorovich Ogarkov would later serve as Professor in the Academy.[7] After graduation in 1948, Vsevolod Ogarkov joined the Soviet Army's BW facility in Kirov, the Institute of Epidemiology and Hygiene. Here he rose to become a head of department and deputy scientific director and is reported to have worked on aerosol and mass immunization technology. One research scientist who grew up in Kirov informed the author that he and his father—also a military biologist—used to join Ogarkov fishing on the local Vyatka river. Later, Ogarkov was to be transferred to the Scientific-Research Institute of Hygiene in Sverdlovsk where he was initially a head of department prior to eventually becoming director. He was awarded his Candidate of Medical Sciences degree in 1954 and his Doctor of Medical Sciences degree in 1963. He became a professor in 1965. His research in Sverdlovsk was focused on the development of an aerosol vaccine against pulmonary forms of plague.[8] Ogarkov's long-term role in senior positions within the USSR Ministry of Defence's BW programme, his party membership since 1949, together with his scientific work on aerosols of pneumonic plague, will have all contributed to his appointment in 1973 as head of the newly formed *Biopreparat*. His position was confirmed by a decree issued by the CPSU Central Committee and USSR Council of Ministers which was signed by Kosygin.[9] It is interesting that precisely at this time rumours began to spread among staff members working on the USSR Ministry of Defence's BW programme about the creation of a special administration within *Glavmikrobioprom*, which would focus on guaranteeing the biological security of the USSR. It was widely understood that officers from the military BW network would be transferred to key positions within the new department.[10]

Such was the power that Ogarkov came to exert on the *Biopreparat* network that it was often referred to as '*Sistema Ogarkova*' or *Ogarkov's System*.[11] However, a number of sources consulted by Zilinskas and Leitenberg describe him as colourless, unimaginative and uninspiring and not a good leader.[12] Alibek supports this rather negative portrait, stating that he was "*uninspiring but pleasant*".[13] Vorob'ev, his immediate deputy, on the other hand, describes him as a "*broad-minded charming person*".[14] Domaradskii reports that he "*had a considerable reputation as a scientist in the 'open' scientific community*".[15] Meanwhile, an international perspective is provided by Reg Bottomley, a consultant to a UK-based fermenter company, who describes an encounter with Ogarkov which took place at the *Biokhimiya* Exhibition in Moscow in June–July 1984. Ogarkov, by this time merely a deputy head of *Glavmikrobioprom*, proceeded immediately to the UK company's stand where air-lift fermenters were on display. He displayed enormous interest in the installations, "*clambered all over the stand*" and asked a number of penetrating, technical questions. At a follow-up meeting in Moscow in September of that year, Ogarkov pressed the UK company for full quotations for complete fermentation systems, including installation costs. He was very anxious to purchase the state-of-the-art systems. Reading the visit report, one is left with the impression of a dynamic individual, held in great respect by those around him.[16]

In 1975 Ogarkov co-authored and published a book with Konstantin Georgievich Gapochko on *Aerogenic Infections* (*Aerogennaya infektsiya*). According to Domaradskii, this book was considered a breach of security by the *Biopreparat* secret service because it revealed "*far too much about how infections are spread through the air or transmitted directly from person to person via the respiratory route*".[17] Ogarkov subsequently received a reprimand from Leonid Brezhnev, the then General Secretary of the Central Committee of the CPSU. This incident appears to have damaged

Ogarkov's reputation among the Soviet leadership, and after subsequent political intrigues, he was relieved of his position in 1979. He did, however, retain his post as deputy head of *Glavmikrobioprom* and was appointed Director of the *glavk*'s All-Union Scientific-Research Institute of Biotechnology—the renamed *VNIIbakpreparat* (see previous chapter).[18]

Ogarkov's immediate deputy at *Biopreparat* was Colonel (Reserves) Anatolii Anatol'evich Miklashevskii—known familiarly to his colleagues as "Tol'ya". This was a highly significant appointment since the latter had previously headed the USSR Ministry of Defence's Scientific-Research Technical Bureau (*NITB*—see above), which up until this point had been directly responsible for the design and maintenance of the Soviet Union's network of BW mobilization preparedness facilities. This was a role that would now be assumed by *Biopreparat*, so Miklashevskii, with his close familiarity with the mobilization capacity, must have been considered to be the perfect appointment. Born in 1920, he is reported to have been well-loved among his colleagues with grey hair, grey moustache, sparkling eyes and infectious laughter. He had served throughout the Second World War on the front-line as a chief of regimental intelligence. He suffered an injury at some point which resulted in one arm being left with limited mobility and being shorter than the other. His personal interests included volleyball, a sport in which he served as a high-level referee. He is described by his colleagues as being intelligent, decisive and energetic. After Ogarkov's dismissal, Colonel Miklashevskii served as director of *Biopreparat*.[19] His role as head though was a very short one, for, in June 1979, he died suddenly of a heart attack.[20] Miklashevskii is reported to have been a friend of the famous Soviet sculptor Georgii Vasil'evich Neroda who, after his death, sculpted a small memorial in his honour at Moscow's Preobrazhenskoe Cemetery. Such was the esteem in which he was held that for more than a decade afterwards, his colleagues would gather at the cemetery on the anniversary of his death.[21]

One of the most high-profile transfers from the USSR Ministry of Defence to *Biopreparat* was that of Anatolii Andreevich Vorob'ev. Born on 5 February 1923 in Kugannyi (Krasnodar *krai*), he had been a student at the Naval Medical Academy (*VMMA*) in Leningrad. In 1945, he joined the Soviet Northern Fleet as a naval doctor. Three years later he re-joined the *VMMA* and initially served as a postgraduate in its Department of Physical and Colloidal Chemistry. Later, in 1951, he obtained his Candidate's degree and was promoted to the position of senior scientific worker. In 1956, he was transferred to the USSR Ministry of Defence's

Scientific-Research Sanitary Institute at *67-i km* settlement, near Zagorsk. Here he was head of a large research department where he was initially focused on research on anatoxins in collaboration with Grigorii Vasil'evich Vygodchikov of the N.F. Gamaleya Institute of Epidemiology and Microbiology (Moscow). Vorob'ev defended his Doctoral dissertation on the construction of sorbed anatoxins. Vorob'ev subsequently collaborated with A.K. Markovich in work to incorporate botulinum toxoids within the *NIISI* Polyvaccine, a standard vaccine for Soviet service personnel.[22] Vorob'ev was also engaged in work on the creation of polyanatoxins including a penta-anatoxin and sexta-anatoxin for the prophylaxis of infections in wounds and botulism.[23] Technology for delivering typhoid polyanatoxin vaccine, botulism toxoid and/or tetanus toxoid via jet injection was developed and successfully trialled in the early 1970s.[24] Lev Fedorov alleges that Vorob'ev was also engaged in research on the combat potential of numerous bacterial toxins and that the production of botulinum and tetanus toxins was reportedly organized.[25]

Another strand of Vorob'ev's work at *67-i km* focused on the study of a number of new "exotic" viruses including Venezuelan, and Western and Eastern Equine Encephalomyelitis. At the beginning of the 1960s he led a military research group which succeeded in developing technologies for the production of immunoglobulins for the treatment and prophylaxis of more than ten viral infections. Vorob'ev also produced an attenuated vaccine strain of the Venezuelan Equine Encephalomyelitis virus and created a live vaccine on its basis. Later, Vorob'ev led research at *67-i km* on the development of new orally administered vaccines. Using original technology, he succeeded in developing peroral preparations against smallpox, plague, Q Fever and Venezuelan Equine Encephalomyelitis.[26]

In 1975, Vorob'ev was promoted to the rank of Major General and three years later he was appointed as deputy scientific director of the Scientific-Research Sanitary Institute. It was in the same year, 1978, that he was to take up the position of first deputy head of *Biopreparat*'s Main Administration for Scientific Work (*Glavnaya upravleniya po nauchnoi rabote*).[27] In his new position, Vorob'ev is reported to have clashed on numerous occasions with Igor' Domaradskii, a senior *Biopreparat* scientist (see above). According to Domaradskii, "*I resented his insatiable ambition and his boastful promises 'to solve all of our complex problems in two or three years'....He seemed principally concerned with enhancing his reputation in his new place of work*".[28] Kanatzhan Baizakovich Alibekov, who replaced Vorob'ev a decade later, is more complimentary and describes him as a "*dignified elderly scientist*".[29]

Another deputy director under Ogarkov, also recruited from the Soviet military, was Colonel (Reserves) Viktor Grigor'evich Popov. Only scant information is known about the background of this individual although he is known originally to have graduated from the highly prestigious Moscow Physical-Technical Institute. He is reported to have headed *Biopreparat*'s technical sciences section and to have had a particular interest in immunology, a field he wished to promote within his new organization. He subsequently coordinated *Biopreparat*'s interactions with Zav'yalov and the Institute of Immunology (see below).[30]

Although, the most prominent, Ogarkov, Miklashevskii, Vorob'ev and Popov were not the only military microbiologists to be transferred to *Biopreparat* in the initial phase of its existence. On his appointment as first deputy, Anatolii Anatol'evich Miklashevskii was given the responsibility to recruit personnel for the association. Given his military background and the aims and objectives of the new association, it is not surprising to learn that he appointed some ten officers or so from the military's BW network to the leading positions in the organization. One of these was Vladimir Dmitrievich Zybin, formerly based at the USSR Ministry of Defence's Scientific-Research Sanitary Institute (Zagorsk), who was appointed head of the Production Department at *Biopreparat*.[31] Another individual to be appointed from the Zagorsk institute was Colonel (Reserves) Lev Aleksandrovich Klyucherev (see above). He took up the position of head of the scientific department at *Biopreparat*.[32]

Biopreparat's Initial Network of R&D and Production Facilities

Under the terms of *Glavmikrobioprom* Order No. 200, issued on 26 June 1974, a number of the Soviet Union's existing dual-use BW facilities were transferred to *Biopreparat*. These included the Berdsk Chemical Factory, the Omutninsk Chemical Factory and the *Progress* Factory (Stepnogorsk, Kazakhstan).[33] It is not known if the dual-use antibiotics facilities at Penza and Kurgan were also transferred at this time, but they were certainly later to come under the control of *Biopreparat*. Part of the Vilnius Experimental-Industrial Factory of Enzyme Preparations was also brought under the control of the association at this time. A large laboratory forming part of the factory was used as the basis for the creation of a major new R&D facility in Lithuania, the All-Union Scientific-Research Institute of Applied Enzymology (Vilnius). This was a facility which was to become of critical strategic importance as one of the only large-scale industrial producers of restriction enzymes used in *Biopreparat*'s research in molecular biology (see below).

The one glaring anomaly in the listing of industrial facilities transferred to *Biopreparat* at this time is the Vyshnii Volochek Factory of Enzyme Preparations located in the Zelenogorskii settlement, 119 kilometres north-west of the Russian city of Kalinin. It has no known links to any military research programmes and has never been referenced with regard to the Soviet offensive BW programme. The plant has its origins in 1889 when it began the production of highly refined starch from imported Indian rice, under the "White Elephant" trademark, which became familiar across Europe.[34] It was initially transferred to the control of *Glavmikrobioprom* in 1967 and after reconstruction is reported to have become one of the leading enterprises of the Soviet enzyme industry.[35]

The only known international interaction with the Vyshnii Volochek Factory occurred in December 1992 when Dr Michael Geisow of the small UK company, Biodigm, paid a visit. He travelled more than 550 km south by road from the city of St Petersburg. He reported that he had been terrified by the drive and that the car had come off the road on two separate occasions.[36] The director at the time, Vladimir Ivanovich Grebenkov, met him upon arrival and explained that the plant occupied an area of 36 hectares and had until recently employed 1300 personnel. The plant had been engaged in the production of human cell and plasma products, including human leukocyte interferon and interleukin—under the brand name *Interlok*. Enzymes had also been produced for a range of uses including as animal feed additives and for brewing. Touring the plant in the period after the collapse of the USSR, Geisow had the impression of dilapidation and decay. However, he did note the existence of a sterile filling plant and a fairly good pilot plant with good autoclaves, spray dryers and so on. He also observed an impressive submerged fungal cultures production line which utilized a continuous train of wagons bearing culture plates for the continuous fermentation of *Aspergillus oryzii* and similar.[37] Even with Geisow's highly informative visit, we have no real idea why this particular plant should have been one of the very few initially selected for inclusion into the newly created *Biopreparat*. It may very well have had a wholly civil function within the organization but without wider access to the relevant archives we can just not be sure.

A number of immensely powerful new R&D facilities were also to be created within the *Biopreparat* network. *Glavmikrobioprom*'s Order No. 200 refers to the transfer to the association of two research institutes at that time being newly organized, the All-Union Scientific-Research Institute of Highly Pure Biopreparations (Leningrad) and the Moscow-based All-Union Scientific-Research Design Institute of Applied Biochemistry (*Vsesoyuznyi nauchno-issledovatel'skii proektno-konstruktorskii institut prikladnoi biokhimii* or *VNIIbiokhimmashproekt*), together with the Berdsk-based Special Design Bureau of Biologically Active Substances (*Spetsial'noe konstruktorsko-tekhnologicheskoe byuro biologiheski aktivnykh veshchestv* or *SKTB BAV*).[38] SKTB BAV—later renamed *NIKTI BAV* (see above)—was the only pre-existing institute to be transferred to *Biopreparat* upon its creation. This appears to make good sense given that its focus was on the supply of a large range of biochemical and microbiological reagents necessary for research in molecular biology.[39] However, Leitenberg and Zilinskas describe the institute as a *Biopreparat* production plant equipped with six 20,000-cubic-metre fermenters. They also report a personal communication from Pasechnik, a leading *Biopreparat* researcher who defected to the UK, suggesting that the institute was known to be in a state of permanent readiness for the production of BW agents including *Francisella tularensis*, *Burkholderia mallei*, brucellae and *Yersinia pestis*.[40] This is the only known reference to *SKTB BAV/NIKTI BAV* having played such a direct role in the Soviet offensive BW programme with Alibek merely alluding to its capabilities regarding the development of restriction enzymes for *Biopreparat* research programmes. It appears a possibility that the facility was confused by Pasechnik with either the Berdsk Chemical Factory itself or the Berdsk Scientific Experimental-Industrial Base (see below).

VNIIbiokhimmashproekt: The Beating Heart of *Biopreparat*

The most prominent component of the newly created *Biopreparat* network was the All-Union Scientific-Research Design Institute of Applied Biochemistry (*VNIIbiokhimmashproekt*). It was engaged in a number of crucial areas including the design and production of fermenters and bioreactors; the production of equipment for microfiltration, sublimation and spray drying; the manufacture of automatic equipment for continuous sterilization and hermetic separation; development of technologies for environmental protection; systems to clean waste gas from microbial processes; and systems to ensure the safety of personnel and protection of the environment when working with dangerous microorganisms.[41] Alibek notes meanwhile, that the institute was engaged in the design of BW manufacturing and testing equipment. He also states it was involved in developing standards for the industrial-scale manufacture of biological weapons.[42] The institute is reported to have played a key role in the engineering of the Berdsk, Omutninsk and Stepnogorsk factories.[43] It was also to design *Biopreparat*'s major R&D centres at Obolensk and Kol'tsovo (see below). Later it was to become the lead facility in the Science Production Association *Biomash*, which itself supplied fermenter installations and other technology to dual-use factories located at Vol'ginskii, Omutninsk and Stepnogorsk.[44]

Given its areas of expertise it is perhaps not surprising to learn that *VNIIbiokhimmashproekt* was the *Biopreparat* facility with the most immediate linkages to the USSR Ministry of Defence. It was intimately intertwined with the USSR Ministry of Defence's *NITB*, which as we have seen, had up to now, had the main responsibility for the design, commissioning and maintenance of the Soviet reserve BW mobilization network. It is therefore not surprising to learn that of the 16 initial core staff members at *VNIIbiokhimmashproekt* no less than nine of the employees had previously worked in *NITB*. Nine of the 16 were graduates of the Military Academy of Chemical Defence, three had graduated from other military academies and only five from civil universities. Nine of the new employees had a Candidates of Technical Sciences degree. Four members of staff at this time had fought in the Second World War and only one person was not a member of the CPSU. The average age of these 16 "*founding fathers*" was 44 and among them only one was a woman.[45]

On 1 April 1974, the minister of defence issued an order regarding the transfer of a group of military officers from *NITB* to *VNIIbiokhimmashproekt*. Nine former *NITB* personnel arrived at the *Biopreparat* building on the specified day to form the core staff of the new institute. One officer, Mikhail Sergeevich Molchanov, was ill and arrived at a later date. Major General (Reserves) Kirill Petrovich Gaponov, the former Chief Engineer of *NITB*, had been appointed director.[46] He had served as a front-line soldier and graduated from the Military Academy of

Chemical Defence. Gaponov appears to have been a modest man, and unlike many of his contemporaries upon reaching an elevated position, he acquired neither his own personal car nor his own summer residence. His deputy and the head of the Scientific-Technological Department was Rustam Khabibovich Sagirov. Another front-line soldier, Sagirov too was a graduate of the Chemical Defence Academy. He later specialized in the freeze-drying of biological preparations. The deputy scientific director was Semen Efimovich Strogov, who in his spare time painted landscapes and still lifes.[47] Another graduate of the Chemical Defence Academy, he had served for three to four years at the Kirov military BW facility. Later he had been appointed head of the Department of Viral Preparations Technology at *NITB*. The chief engineer of the institute was Lev Georgievich Vlasko. He had also served as a front-line soldier and been promoted to the rank of officer. He later worked for a decade in the military BW facility in Sverdlovsk and was then employed alongside his wife in *NITB*.[48]

Another *NITB* staff member to join *VNIIbiokhimmashproekt* was Vitalii Andreevich Grishin. Born in 1938 in Orel, he had graduated from the Chemical Defence Academy. Here he had become close personal friends with his classmate Yurii Tikhonovich Kalinin, the future head of *Biopreparat*. After military service in Germany, he joined *NITB* in 1964. Here he specialised in the selection, checking, installation and operation of fine air filters. Part of his responsibilities involved visiting facilities to test installed filtration systems and the training of staff in the use of them. Another staff member was Mikhail Luk'yanovich Davydov who headed the institute's Department of Packaging Technology for Special Products. Nothing is known of his background.[49] Aleksei Ivanovich Izvol'skii was another front-line soldier who, during the Second World War, fought all the way to Berlin. He then graduated from the Chemical Defence Academy and served in facilities subordinate to the Soviet chemical troops. He was the only non-member of the CPSU and was appointed as head of *VNIIbiokhimmashproekt*'s Department of Technology for Packaging and Storage of Biological Products.[50] Valerii Pavlovich Mityushev, an electrical engineer, joined the institute from *NITB*. He was appointed as head of the Department of Control and Automation of Technological Processes.[51]

Mikhail Mikhailovich Molokoedov was appointed as leading designer within *VNIIbiokhimmashproekt*'s Dosing Equipment Development Department. He was actually the son of Mikhail Markelovich Molokoedov, who himself had worked for many years in the military's BW facility at Sverdlovsk. Born in 1943 he was unusual in that he had graduated from the Moscow Aviation Institute and had later been employed in an aviation design bureau. He then joined *NITB* around 1970.[52] Another appointee at *VNIIbiokhimmashproekt* was Mikhail Sergeevich Molchanov. Born in 1929 in Tula, he initially served as commander in a chemical troops platoon. Later he enrolled in the Chemical Defence Academy, and then joined *NITB* where he became an expert in the installation of equipment, carrying out many inspections to guarantee the quality of work. He later headed up the Installation Department at *VNIIbiokhimmashproekt*.[53] Meanwhile, Anatolii Yakovlevich Naidenov was appointed head of the new institute's Department of Special Safety Technology. He was born in 1931 and had graduated from the Military Engineering Academy. Prior to joining *NITB*, he had served at the military BW facility in Kirov.[54]

Vladimir Sergeevich Redin was appointed the head of *VNIIbiokhimmashproekt*'s Design Department for the Development of Capacitive Equipment. He had graduated from the Military Engineering Academy and then joined the military BW facility in Kirov. His main scientific interest was focused on mixing devices in bioreactors. He later joined *NITB*.[55] The only female scientist to be in the initial group joining *VNIIbiokhimmashproekt* was Evdokia Afanaseevna Rudakova. She was born in 1928 and graduated from the Moscow Institute of Chemical Engineering. Later she worked in the military BW facility in Sverdlovsk. Upon being transferred to *VNIIbiokhimmashproekt*, she was appointed as head of the Department for the Development of the Structure and Layout of Microbiological Laboratories. It was in this capacity that she played a lead role in the design of *Biopreparat*'s two largest R&D centres in Obolensk and Kol'tsovo.[56] A colleague of hers in the new facility was Boris Viktorovich Sergeev. He was head of the Department of Complex Problems. He was born in 1934 and had graduated from the Chemical Defence Academy.[57]

Another graduate from the Chemical Defence Academy was Lev Petrovich Shevelev. Born in 1931, after graduation he served in the armed forces in Hungary and then joined *NITB*. At *VNIIbiokhimmashproekt* he was appointed as head of the Cultivation Department. He later secured an appointment higher-up in the *Biopreparat* management structure. A senior researcher and head of the Special Safety Engineering Department at the new institute was Anatolii Pavlovich Shmelev. Born in 1931 he arrived as a senior technician in the military BW facility in Zagorsk in 1955. He served for 10–12 years at this location prior to being transferred for 2–3 years to the Vozrozhdenie Island proving ground. He was then transferred to *NITB*.[58] Finally, there was Arutyun Vartanovich Eiromdzhants who headed *VNIIbiokhimmashproekt*'s Design Department for the Development of Dosing Devices. He was born in 1930 and in 1955 or 1956 joined the BW facility in Zagorsk as a senior technician in the design department. Later he was transferred to its military sister facility in Sverdlovsk. He returned to Moscow in the early 1960s.[59]

At this early stage of its existence, *VNIIbiokhimmashproekt* comprised the 17 former military personnel and not much else. It possessed no office equipment nor its own accommodation and was only allocated a small office with two tables and a telephone in the *Biopreparat* HQ building itself. The salaries of the staff and any other financial transactions of the institute were handled by the *Biopreparat* accounting department.[60] Under the terms of the government order, it was envisaged that a special complex of buildings would be constructed to accommodate the institute. However, at this stage it was assumed that it would rent premises somewhere in Moscow. This proved a most difficult task since, besides rent, the new institute could offer potential landlords no barter of products or manufactured goods. Staff at *VNIIbiokhimmashproekt* are reported to have become under immediate pressure from Belyaev, head of *Glavmikrobioprom*, for resolving technical issues associated with safety of personnel and protection of the environment at a new facility being constructed at a second site in Stepnogorsk—the dedicated BW facility known as the Scientific Experimental-Industrial Base (*SNOPB*). Belyaev is reported to have personally monitored progress on this project and to have made the final decision on the institute's proposals. Regular meetings were held with him in the early evenings at *Glavmikrobioprom*'s HQ on ulitsa Lesteva. Again, this points most strongly to the special importance attached to this particular project by the Soviet leadership.[61]

VNIIbiokhimmashproekt finally managed to rent the upper fifth floor of a school building which was the first legal address of the new institute. One of its first actions was to form a primary party organization which was registered with the Kuibyshev District Committee (*raikom*) of the CPSU. When, after a few months, the school requested the institute to vacate its building, the raikom assisted it in moving to the top floor of a local telephone exchange on Baikal'skaya ulitsa in Moscow's Gol'yanovo district. *VNIIbiokhimmashproekt* at this time also rented a church which was used as a warehouse to store some newly arrived equipment and laboratory furniture. It was at this point that the unit began to resemble a real, functioning institute, with security and a pass system in place, and the allocation of an office to the director, with a secretary in reception.[62]

In 1977, *VNIIbiokhimmashproekt* succeeded in renting an entire school building on Malaya Filevskaya ulitsa close to the Pionerskaya metro station in the south-western outskirts of Moscow. Part of the institute's staff were transferred to the new site with the remainder staying put in the Gol'yanovo district. Meanwhile, construction had finally been initiated of *VNIIbiokhimmashproekt*'s new dedicated building in north-west Moscow at ulitsa Klary Tsetkin, 4, close to the Voykovskaya metro station. The construction project immediately ran into bureaucratic problems and the institute's management decided these could only be resolved by scheduling a meeting with Viktor Vasil'evich Grishin, the First Secretary of the Moscow City Committee of the CPSU and a member of the Politburo. He was apparently, the only member of the city leadership who had had direct access to the full text of the secret resolution of the Central Committee of the CPSU and Council of Ministers regarding the creation of the institute. After a series of attempts, Gaponov finally managed to schedule a meeting with Grishin who made a telephone call to an official whom he instructed to ensure the execution of the resolution.[63]

The design of the new complex on ulitsa Klary Tsetkin was carried out by *VNIIbiokhimmashproekt*'s own staff members. In order to accelerate the project, they used a typical three storey school as the base design for the first two buildings, A and B—connected by overhead walkways, and then added an extra storey to each and made alterations to the internal layouts. A computing centre was installed on the second floor of Building A along with four rooms allocated to the Department of Control and Automation of Technological Processes. By the end of 1978, Building A had become operational. Early in 1980, the rented buildings were evacuated and staff moved into the new premises. It was envisaged that Building B would house a conference hall, exhibition hall, a dining area and administrative area. It was at this point that construction of the complexes main 11-storey Building "G", which would eventually house the institute's laboratories, began. Only the design unit remained in rented accommodation in a building near Moscow's Tishinsk market. It was envisaged that it would eventually move into Buildings A and B after the transfer of the laboratories to the completed Building G.[64]

In addition to the institute itself, a 12-storey residential building for staff and their families was constructed a mere 10 minutes' walk from the institute. Staff were also provided with privileged access to a special network of polyclinics, hospitals and sanatoria run by the USSR Ministry of Health's Fourth Main Directorate. In 1983, a medical unit was built for the institute close to the Kolomoisky metro station. It was also used by other employees within the *Biopreparat* and *Glavmikrobioprom* networks. The level of service provided by this closed medical institution is reported to have been significantly higher than that offered by regular clinics in the Soviet public health system.[65]

One of the very first new members of staff, outside of the core military group, to be appointed to *VNIIbiokhimmashproekt* was Galina Ivanovna Agalakova. She arrived shortly after the new institute was opened and took up the position of head of the personnel department. She was to hold this position for more than a quarter of a century, and at the height of the *Biopreparat* programme built-up staff levels to 1500 personnel. She was aided in her work by Valerii Kuzmich Zemtsov, the deputy head of security at the institute.[66] This was a key position since employees of *VNIIbiokhimmashproekt* potentially had access to the entire *Biopreparat* network including its BW mobilization capacity. Once again, former employees of the USSR Ministry of Defence were targeted for recruitment. Valerii Antonovich Ambrosov, for example, was recruited from the military's BW facility in Zagorsk. He was a Doctor of Biological Sciences and initially headed the Fermentation Department's Laboratory of Cell Cultures. Subsequently, the lab was spun-off to form its own independent department. Later Ambrosov was transferred to the Central Committee of the CPSU where his work focused on Soviet biological programmes. As far as the institute was concerned, the move could have potential benefits since it opened up informal channels to the higher party echelons. Another Zagorsk scientist to be recruited was Anatolii Dmitrievich Ukraintsev. He was a candidate of Medical Sciences and replaced Ambrosov as head of department before eventually rising to the position of director. Another key defence scientist to be recruited was Salikh Salakhovich Muratov, who had formerly been based at the military's Kirov BW facility.[67]

One interesting aspect of the *VNIIbiokhimmashproekt* recruitment programme is the fact that it also targeted so-called second-generation scientists—individuals who had grown up as children at Soviet military BW bases and then followed in their parent's footsteps to become military biologists in their own right. Another group to be targeted for recruitment were students at a Moscow food institute. *VNIIbiokhimmashproekt*'s director, Gaponov, was a teacher in biotechnology at the institute and he selected talented individuals to work in the fermentation and other departments in his facility.[68]

VNIIbiokhimmashproekt's Affiliated Sister Institutes

VNIIbiokhimmashproekt was very closely linked to two new facilities which were to be created by *Biopreparat* for the design and production of specialized instruments and equipment, the Experimental Design Bureau of Precision Biological Engineering (*Opytno-konstruktorskoe byuro tonkogo biologicheskogo mashinostroeniya* or *OKB TBM*) in Kirishi and the Experimental Design Bureau of Automation Equipment (*Opytno-konstruktorskoe byuro sredstv avtomatizatsii* or *OKBA*) in Yoshkar-Ola. The Kirishi design institute was subsequently transferred to the control of the Leningrad-based All-Union Scientific-Research Institute of Highly Pure Biopreparations (*VNIIOChB*). However, despite this organizational switch, the ties to *VNIIbiokhimmashproekt* remained close and the institutions pursued a number of joint projects.

As already indicated, *OKB TM* was located in the town of Kirishi situated on the Vol'khov River some 115 kilometres southeast of Leningrad. The site chosen for the creation of the bureau may have been influenced by the fact that *Glavmikrobioprom* was already constructing a plant in the town with anticipated capacity for the production of 100,000 tonnes of petroleum-based microbial protein. Some engineering facilities may already have sprung up to service the new plant. Whatever the case, planning of the construction of *OKB TM* was initiated in December 1973 when Vladimir Dmitrievich Zybin together with Valerii Mityushev from the USSR Ministry of Defence's Scientific-Research Technical Bureau travelled to Leningrad to meet with the design organization, *Gipropribor*, which had responsibility for construction of the new facility.[69]

Vladimir Pasechnik (see below), the former Director of *VNIIOChB* (Leningrad) who defected to the UK, reported that *OKB TM* was engaged in work to develop new equipment both for the production and delivery of biological weapons.[70] These included large-scale fermenters and filtration, separation and chromatographic equipment together with explosion chambers. Fermenter installations and other technology were subsequently supplied to various facilities within the *Biopreparat* network including its R&D centre at Obolensk (Serpukhov *raion*, Moscow *oblast'*). Some additional information about the Kirishi facility was also made available to the author when, in May 1994, in the wake of the collapse of the

USSR, he visited *OKB TM*. It was at this time found to be occupying a shared site with the Factory of Biological Engineering (*Kirishskii zavod biologicheskogo mashinostroeniya* or *KZBM*), with both units being protected by a six-feet high eight-stranded electrified fence. It is not known how long at this point *KZBM* had had its own independent existence but certainly the impression was that in reality the two units had formed, until very recently, a single R&D and manufacturing complex. The factory occupied a four-storied building approximately 150 metres in length. Two smaller multi-storied buildings were identified as belonging to *OKB TM*. These were each connected by overhead walkways to *KZBM*. By this time, with a new Russian government in place, there was no visible sign of any strong security presence at the main pedestrian entrance to the facilities which was controlled by a turnstile-type device. However, inside the main production area, where equipment was manufactured and stored, virtually every door had a push-button entry system or was heavily padlocked.[71] The complex is reported to have employed 2000 personnel at its height but following the collapse of the USSR had been forced to scale down its workforce to 700 people.[72]

The design and development of fermenters was being undertaken in the Experimental-Design Bureau of Precision Biological Engineering. Extensive drawing facilities for design work were in evidence. High-quality control equipment and circuitry had been developed here as well as excellent magnetic drive couplings on fermenters. The vast building occupied by the Factory of Biological Engineering meanwhile possessed machine shops and fabrication shops. Several store rooms were packed with spare parts including electric motors for pumps. A number of fermenters, fermentation installations and other equipment were in the final stages of preparation for shipment to unknown locations. These included: a complete 100-litre fermentation installation which was to be supplied to a factory; one 1000-litre fermenter; four 500-litre fermenters; four filtration and concentration systems complete with filter presses and ultrafiltration; one 100-litre chromatography column and various smaller ones; assorted fraction collectors; and one very heavily built Class III glove box (with very high containment) fabricated from stainless steel. The maximum volume of fermenters manufactured at the complex was reported to be 10,000 litres. The factory at this time had also sought to diversify its output and had completed production of around 30 brightly painted invalid tricycles.[73]

OKB TM was complimented in its work by the Experimental Design Bureau of Automation Equipment (*Opytno-konstruktorskoe byuro sredstv avtomatizatsii* or *OKBA*) in the old fortress city of Yoshkar-Ola, located in the remote Mari El Republic. Again, planning of the construction of *OKBA* was initiated in December 1973, following a meeting in Leningrad between *Gipropribor*, which had responsibility for construction of the new facility, with Vladimir Dmitrievich Zybin and Valerii Mityushev (from the USSR Ministry of Defence's Scientific-Research Technical Bureau). The only information available with regard to activities at *OKBA* is provided by Pasechnik who points to its important role in the manufacture of small bomblets and filling stations for the automatic filling of warhead devices with infectious agents. It was also engaged in the manufacture of fermenters, gas and liquid chromatographs, bacterial colony counters, disintegration machines and freeze driers.[74]

The choice of location for the creation of *OKBA* is again an interesting one and may conceivably be connected to plans which later emerged for the creation of a major new offensive BW facility in the Mari El Republic. The design bureau was in fact located on Ulitsa Stroitelei, 94 in Yoshkar-Ola which is very close to the site of the Yoshkar-Ola Vitamin Factory on nearby ulitsa Karla Marksa. The latter plant had been created on 10 March 1942 by the People's Commissariat of the Food Industry to supply vitamin products to the Red Army. Presumably, it was later absorbed by *Glavmikrobioprom*. It subsequently was to be used as the basis for the planned creation of a viral-based BW production plant. Alibek reports that in 1985, Mikhail Gorbachev, then CPSU First Secretary, signed-off a Five-Year Plan relating to the ambitious expansion of the Soviet BW programme. It included the allocation of 300 million roubles (around US$400 million) for the construction by *Biopreparat* of a viral production plant in the Yoshkar-Ola region.[75] Ostensibly, the new facility, which was to be constructed at a remote site in Mochalishche, around 46 kilometres to the south-east of Yoshkar-Ola, was to be a branch of the vitamin factory. However, its anticipated production capacity was to be six times that of its parent facility. Its intended capabilities with regard to biological weapons agents are unknown.

Order No. 28 issued by the USSR Ministry of the Medical Industry (see below) on 26 January 1990, specifically refers to the production of an influenza (viral) vaccine, nutrient media and dry ginseng biomass at the vitamin factory's Mochalishche site. *VNIIbiokhimmashproekt* (see above) was listed as a key participant in the project. Crucially, the implementation of the order was entrusted to Yurii Kalinin, head of *Biopreparat*.[76] The project was so highly classified that nobody in the republic was made aware of its true purpose including the first secretary of the republic's CPSU committee. Construction was initiated at the site in 1987. However, it soon became the target of a vociferous local popular campaign directed against its potential environmental impact including fears concerning the contamination of a tributary of the Volga and the threat posed by atmospheric emissions. A special commission was subsequently appointed in response to the campaign and declared the project unfeasible. It was as a result mothballed and later abandoned.[77] Leitenberg and Zilinskas point to a lack of funds as being key in the decision to halt construction at the site.[78]

Another design institute which developed very close connections with *VNIIbiokhimmashproekt* was the Moscow branch of the State Institute for the Design of Enterprises of the Hydrolysis Industry and Microbiological Synthesis—known as Mosgiprobiosintez.[79] The Russian State Archive of the Economy suggests that the latter institute was officially transferred to *Biopreparat* under the terms of Order No. 389 issued on 28 December 1976.[80] However, in reality, by 1974 it had already been absorbed by the association and was sharing accommodation with *VNIIbiokhimmashproekt* in an old house located close to Moscow's Kholkhoznaya metro station.[81] Possibly *Mosgiprobiosintez* was shortly thereafter renamed *Giprobioprom*—the State Institute for the Design of Enterprises of the Biological Industry, which Alibek reports played an important role in the design of BW research and production facilities.

Another facility, which worked very closely with *VNIIbiokhimmashproekt* was *Biopreparat*'s All-Union Scientific-Research Institute of Biological Instrument Making (*Vsesoyuznyi nauchno-issledovatel'skii institut biologicheskogo priborostroeniya* or *VNII BP*). The first director and founder of the facility was Lieutenant General (Reserves) Yurii Tikhonovich Kalinin, later in 1980 to become the head of the entire *Biopreparat* network.[82] His military background made him highly suitable for the post, having graduated in 1965 from the Higher Military Academy of Chemical Defence and later serving in the USSR Ministry of Defence's Scientific-Research Sanitary Institute (*67-i* settlement, Zagorsk). *VNII BP* was eventually housed at Volokolamskoe shosse, 75 in the north-western outskirts of Moscow. The institute had a strong emphasis on aspects of biosafety, developing a range of systems for monitoring the presence of dangerous pathogens in the air and a portable apparatus for sampling biological aerosols. Under Kalinin's leadership the institute developed the first Soviet manufactured immunoenzyme test systems for the detection of high-risk viral and bacterial pathogens. It is also reported to have developed new methods of microanalysis and instruments based on photo- and chemiluminescence, gas and liquid chromatography, mass spectrometry and chromatography-mass spectrometry. Subsequently, most of this technology was put into serial production in Russian industry, including photometers for monitoring blood sugar level in diabetes patients, Kristall chromatographs for analytical laboratories, and photoluminescent sensors for immunofluorescence analysis.[83]

Mention should be briefly made of one more institution with which *VNIIbiokhimmashproekt* interacted closely. This was the Science Production Association *Promavtomatika* (*Nauchno-proizvodstvennoe ob'edinenie Promavtomatika* or *NPO Promavtomatika*) based in Grozny, the capital of Chechnya. It was subordinate to the USSR Ministry of Instrumentation, Automation and Control Systems (*Ministerstvo priborostroeniya, sredstv avtomatizatsii i system upravleniya* or *Minpribor*). *Promavtomatika* itself comprised of a branch of a Sumgait-based petrochemical research and design automation institute, a design bureau focused on automation and a pilot plant. Given that Grozny was no longer a major centre of oil production and refining, *Promavtomatika*—which had some previous experience in automating an industrial microbiology process for producing acetone—was redirected towards the integration of automated control systems into the microbiological industry. The first contacts between *VNIIbiokhimmashproekt* and the association took place on 1 5 September 1974 in Grozny with Gaponov (director of the former) and Aleksandr Ivanovich Myalkin (head of department at *Biopreparat*) participating in the meetings. *Promavtomatika* was subsequently involved in a number of projects including the development of automated control systems at *Biopreparat*'s Berdsk Chemical Factory (a dual-use facility—see above), and at its BW R&D centre in Obolensk, the All-Union Scientific-Research Institute of Applied Microbiology (*VNII PM*)—see below.[84]

The Creation of the Science Production Association *Biomash* (*NPO Biomash*)

Sometime around the mid-1980s the intimate links between *VNIIbiokhimmashproekt* and *OKBA* (Yoshkar-Ola) were officially formalized with the creation of the Science Production Association *Biomash*. It was headed by *VNIIbiokhimmashproekt*'s director, Gaponov, and shared its headquarters with the institute in Moscow at ulitsa Klary Tsetkin, 4. Besides *OKBA*, the new association also incorporated the Lotoshino Biochemical Factory. Despite the attempt to break down barriers to innovation with this organizational fusion of science and production, the association was deemed a failure and the three subordinate entities continued to prioritize the pursuit of their own independent programmes.[85] Despite this perceived lack of success, the association did continue to survive. During a visit made to *NPO Biomash* by the author in May 1993—shortly after the departure of Alibekov as director, three complete fermentation installations were being assembled which were to be shipped in large transport containers standing nearby. Recent developments at the association at this time included the creation of mobile fermentation units which could be delivered to oil production sites and which would then be used to manufacture special microbial polysaccharides able to increase yields; the development of fermenters for plant cell production, and development of a system for sterile probing during fermentation. *NPO Biomash* reported that it continued at this time to be engaged in the joint development of fermentation reactors with *OKB TM*.[86]

Notes

1. Introduction to the *Glavmikrobioprom* archive, RGAE, Fond 178, Opis 1, undated document.
2. Ibid. *Biopreparat* is the shortened form of "*biologicheskii preparat*", translated as "biological preparation".
3. Aman, R., Cooper, J. (Eds.), *Industrial Innovation in the Soviet Union*, Yale University Press, London, 1982, p. 458.
4. Domaradskii, I., *Znanie-sila*, p. 63.
5. Hoffman, *The Dead Hand*, p. 140.
6. Domaradskij, Orent, *Biowarrior*, p. 154.
7. *Professora voenno-meditsinskoi (mediko-khirurgicheskoi) akademii (1798–1998)*, Nauka, St Petersburg, p. 80.
8. Rimmington, *Who's Who*, p. 211.

9. Vorob'ev, A.A., *Dvadtsatyi vek moimi glazami*, Medical Information Agency, Moscow, 2005, p. 204.
10. Mityushev, *Zapiski obyknovennogo cheloveka*, p. 416.
11. Leskov, S., Chuma i bomba. Voennye bakteriologicheskie programmy v Rossii i SShA razrabatyvayutsya v glubokoi taine i predstvlyayut dlya mira strashnuyu ugrozu, *Izvestiya*, No. 118(23,973), 26 June 1993, p. 15.
12. Leitenberg, Zilinskas, with Kuhn, *The Soviet Biological Weapons Programme*, p. 158.
13. Alibek, with Handelman, *Biohazard*, p. 22.
14. Vorob'ev, A.A., *Dvadtsatyi vek moimi glazami*, p. 204.
15. Domaradskij, Orent, *Biowarrior*, p. 193.
16. Bottomley, R., *Visit Report, USSR*, Peruvum Ltd., 9–14 September 1984.
17. Domaradskij, Orent, *Biowarrior*, p. 193.
18. Rimmington, *Who's Who* (1994), p. 211. There is evidence that, in this position, Ogarkov continued to "hoover-up" defence-related technologies for *Glavmikrobioprom*. For example, records held within the Russian State Archive of the Economy (*RGAE*) reveal that on 28 February 1985, Anatolii Petrovich Aleksandrov, the President of the USSR Academy of Sciences, wrote a letter to *Glavmikrobioprom* concerning the transfer of the Scientific-Research Institute for Biological Testing of Chemical Compounds' Laboratory of Molecular Genetics of Cell Stability (with 20 personnel) from the USSR Ministry of the Medical Industry (*Minmedprom*) to one of the institutes belonging to *Glavmikrobioprom*. The head of the laboratory at this time is recorded as L.L. Gumanov (Candidate of Biological Sciences). According to Aleksandrov, the work of the laboratory "*opens up new possibilities for the theoretical and applied selection of microorganisms and plants and has fundamental significance for agriculture, medicine and defence*". Ogarkov, at this time deputy head of *Glavmikrobioprom*, is recorded as having given his preliminary approval for the transfer to take place. Ogarkov died on 16 August 1987 and is buried in plot 10i in the Kuntsevo Cemetery, Moscow.
19. Mityushev, *Zapiski obyknovennogo cheloveka*, p. 427.
20. Ibid., p. 427.
21. Ibid., p. 364.
22. *Soviet Measures for Biological Warfare Defence*, Ministry of Defence, Defence Intelligence Staff (Directorate of Scientific and Technical Intelligence), 5 August 1968.
23. Anatolii Andreevich Vorob'ev (K 70-letiyu co dnya rozhdeniya), *Voenno-meditsinskii zhurnal*, No. 2, February 1993, pp. 75–76.
24. Agafonov, V.I., Babkin, E.I., Bulatova, T.I., Gamleshko, G.P., Gapochko, K.G. Bezygol'nyi metod immunizatsii assotsiirovannymi sorbirovannymi vaktsinami [Jet method of immunizing with associated adsorbed vac-

cines]. *Voenno-meditsinskii zhurnal*, No. 12, December 1974, pp. 44–48, carried at www.cdc.gov/nip/dev/Jetinject-Bib.doc, accessed on 23 September 2003.
25. Fedorov, L., Death from the Test-Tube, *New Times*, September 1998, pp. 22–26.
26. Anatolii Andreevich Vorob'ev (K 70-letiyu co dnya rozhdeniya), *Voenno-meditsinskii zhurnal*, pp. 75–76.
27. Vorob'ev, *Dvadtsatyi vek moimi glazami*, p. 205.
28. Domaradskij, Orent, *Biowarrior*, p. 191.
29. Alibek, with Handelman, *Biohazard*, p. 117. In March 1988, Vorob'ev represented the USSR Ministry of the Medical and Microbiological Industry (*Minmedbioprom*) at a meeting in Moscow of the UK/USSR Working Group on the Medical Industry and Biotechnology. After departing *Biopreparat*, Vorob'ev continued with his scientific career at the I.M. Sechenov Moscow Medical Academy. In 2002, according to his published memoirs, he visited Professor Christopher Lowe at Cambridge University in the UK. Vorob'ev died on 23 March 2006, an announcement was made on that same day by the Yu.A. Ovchinnikov Society of Russian Biotechnologists, http://www.biorosinfo.ru.
30. Leitenberg, Zilinskas, with Kuhn, *The Soviet Biological Weapons Programme*, p. 263.
31. Ibid., p. 418.
32. Domaradskij, I., Orent, W., *Biowarrior*, p. 199.
33. Introduction to the *Glavmikrobioprom* archive, *RGAE*, Fond 178, Opis 1, undated document.
34. Kompaniya "DEKO"—proizvodstvo fermentnykh preparatov, http://dekopharm.ru/, Accessed on 7 August 2020.
35. Zelenogorskii (Tverskaya oblast'), https://ru.wikipedia.org/wiki/Zelenogorskii_(Tverskaya_oblast'), Accessed on 7 August 2020.
36. Information provided to author by Dr Michael Geisow, 29 June 1994.
37. Geisow, M., *Visit to Vishney Volochek Dec. 1992*, Biodigm, 7 December 1994, p. 2.
38. Introduction to the *Glavmikrobioprom* archive, *RGAE*, Fond 178, Opis 1, undated document.
39. Filial FBUN GNTs VB "Vektor" Institut meditsinskoi biotekhnologii, http://www.vector.nsc.ru/filial-/, Accessed on 12 August 2020.
40. Leitenberg, Zilinskas, with Kuhn, *The Soviet Biological Weapons Programme*, pp. 161–162.
41. Rimmington, A., *The Biotechnology & Pharmaceutical Industry in Russia 2010: A Comprehensive Guide, Part I: R&D Centres and Institutes*, Birmingham, October 2010, p. 12; Mityushev, *Zapiski obyknovennogo cheloveka*, p. 422.

42. Alibek, with Handelman, *Biohazard*, p. 300.
43. Leitenberg, Zilinskas, with Kuhn, *The Soviet Biological Weapons Programme*, p. 163.
44. Information supplied by representative of Science Production Association *Biomash*, ulitsa Klary Tsetkin 4/6, Moscow, 31 May 1993.
45. Ibid., p. 430.
46. Mityushev, *Zapiski obyknovennogo cheloveka*, p. 417.
47. Ibid., p. 434.
48. Ibid., p. 432.
49. Ibid., p. 436.
50. Ibid., p. 437.
51. Ibid., p. 438.
52. Ibid., p. 439.
53. Ibid., p. 440.
54. Ibid., p. 441.
55. Ibid., p. 442.
56. Ibid., p. 443.
57. Ibid., p. 444.
58. Ibid., p. 446.
59. Ibid., p. 447.
60. Ibid., p. 421.
61. Ibid., p. 422.
62. Ibid., p. 424.
63. Ibid., p. 425.
64. Ibid., p. 428.
65. Ibid., p. 429.
66. Ibid., p. 448.
67. Ibid., pp. 450–451.
68. Ibid., p. 449.
69. Ibid., pp. 418–419.
70. Adams, J., *The New Spies: Exploring the Frontiers of Espionage*, Hutchinson, London, 1994, p. 273.
71. Observations made during the visit to Kirishi by Dr Anthony Rimmington, University of Birmingham on 13 May 1994.
72. Ibid.
73. Ibid.
74. Leitenberg, Zilinskas, with Kuhn, *The Soviet Biological Weapons Programme*, p. 163.
75. Alibek, with Handelman, *Biohazard*, p. 117.
76. Bykov, V.A., O proektirovanii ob"ektov rasshireniya Ioshkar-Olinskogo vitaminogo zavoda v pos. Mochalishche s organizatsiei dopolnitel'nykh moshchnostei po vypusku grippoznoi vaktsiny, sukhikh pitatel'nykh sred i

sukhoi biomassy zhen'shenya, USSR Ministry of the Medical Industry, Order No. 28, 26 January 1990, *RGAE*, Fond 788, Opis 1, No. 1339, p. 85.

77. Fedorov, L., *Soviet Biological Weapons: History. Ecology. Politics*, pp. 127–128.
78. Leitenberg, Zilinskas, with Kuhn, *The Soviet Biological Weapons Programme*, p. 588.
79. Mityushev, *Zapiski obyknovennogo cheloveka*, p. 421.
80. Introduction to the *Glavmikrobioprom* archive, *RGAE*, Fond 178, Opis 1, undated document.
81. Mityushev, *Zapiski obyknovennogo cheloveka*, p. 421.
82. *Gosudarstvennyi nauchnyi tsentr gosudarstvennyi nauchno-issledovatel'skii institut biologicheskogo priborostroeniya: 25 let*, Moscow, 2000, p. 1.
83. *Kalinin, Yurii Tikhonovich*, 2002, Entry at www.biograph.comstar.ru, Accessed on 2 February 2005.
84. Mityushev, *Zapiski obyknovennogo cheloveka*, pp. 561–564.
85. Mityushev, *Zapiski obyknovennogo cheloveka*, p. 484.
86. Information provided by representative of the Science Production Association *Biomash*, ulitsa Klary Tsetkin 4/6, Moscow, 31 May 1993.

CHAPTER 5

Anthrax on the Kazakh Steppe: *Biopreparat*'s Network of Experimental-Industrial Bases

As has already been documented, under the terms of *Glavmikrobioprom* Order No. 200, issued on 26 June 1974, a number of the Soviet Union's existing dual-use BW facilities were transferred to *Biopreparat*. These included the Berdsk Chemical Factory, the Omutninsk Chemical Factory and the *Progress* Factory (Stepnogorsk, Kazakhstan).[1] Construction of the first of these factories—which incorporated BW mobilization facilities—had begun in Berdsk as early as 1958. These large-scale facilities were to comprise *Biopreparat*'s core capability with regard to the production of biological weapons. Some idea of the sheer scale of these factories is evidenced by the numbers of people employed at the sites. By 1986, according to the Soviet archives, some 2038 personnel were employed at the Omutninsk factory, 2688 personnel at the Berdsk factory and 3126 at the *Progress* factory.[2]

Immediately upon its creation, *Biopreparat* took the decision to create new pilot and industrial-scale facilities at the sites of its Berdsk, Omutninsk and Stepnogorsk production plants. The new facilities were initially subordinate to the All-Union Scientific-Research Design Institute of Applied Biochemistry (*VNIIbiokhimmashproekt*) which played a key role in their design and commissioning. Three new entities were created: the Kazakh Branch of the Scientific Research Design Institute of Applied Biochemistry (Stepnogorsk); the Siberian Branch of the Scientific-Research Design Institute of Applied Biochemistry (Berdsk) and the East European Branch

A. Rimmington, *The Soviet Union's Invisible Weapons of Mass Destruction*, https://doi.org/10.1007/978-3-030-82882-0_5

of the Scientific-Research Design Institute of Applied Biochemistry (Omutninsk). It is possible that *Biopreparat* used the public affiliations to *VNIIbiokhimmashproekt* to disguise the intent and purpose of the new facilities which were being constructed, but very soon thereafter they were renamed and reorganized as "experimental-industrial bases" and ceased to report to the applied biochemistry institute.

The Omutninsk Scientific Experimental-Industrial Base (*ONOPB*)

Shortly after the creation of *Biopreparat*, construction began in Vostochnyi (Kirov *oblast'*) of the East European Branch of the Scientific-Research Design Institute of Applied Biochemistry. The new facility, subsequently renamed the Omutninsk Scientific Experimental-Industrial Base (*Omutninskaya nauchnaya opytno-promyshlennaya baza*, abbreviated to *ONOPB*) was built on territory immediately contiguous with the pre-existing Omutninsk Chemical Plant.[3] *ONOPB* was referred to in *Biopreparat*'s coded cables as Post Office Box Number V-8389 (*p/ya V-8389*). Convict labour, supplied by the USSR Ministry of Internal Affairs (*MVD*) under the terms of a secret agreement with *Biopreparat*, was utilized for the construction work at Vostochnyi.[4]

A key project which fell under the auspices of *ONOPB* was the construction of a new biological weapons pilot plant known as Building 107. The new facility was a three-storied structure and was, according to Alibek, utilized as a pilot plant for production of *Francisella tularensis* (the causative agent of tularaemia). It comprised three zones. Zone 1 was the outer shell and incorporated administrative and security personnel and laboratories utilized for non-infectious organisms. Within Zone 1 was Zone 2 which consisted of high-containment laboratories utilized for pathogenic microorganisms, storage vaults, animal holding facilities and 16-tonne and 20-tonne fermenters. In the interior of the building was Zone 3 which contained downstream processing facilities including centrifuges and drying and milling machines. Zones 2 and 3 were in possession of their own air supply and maintained under negative air-pressure, together with hydrogen peroxide being sprayed into the air from nozzles.[5]

Kanatzhan Baizakovich Alibekov is reported to have begun his employment at *ONOPB* in August 1975. He was originally employed as a junior scientist. In March 1976 he was transferred to Berdsk (see below) but returned to *ONOPB* in 1980. By this time, he was a senior lieutenant and took up the position of chief of the Technical Department with responsibility for 150 people. In 1981 he was appointed by Kalinin as deputy director and soon after became acting director of *ONOPB* with responsibility for 400 military personnel.[6] The importance attached to the *ONOPB* project by the Soviet political and military leadership was underlined by an inspection which took place in February 1981 of Building 107 by Nikolai Aleksandrovich Shakhov, head of the Military Department of the Central Committee of the CPSU and Lieutenant-General Lebedinskii (at this time deputy head of the USSR Ministry of Defence's Fifteenth Administration which controlled a network of military BW facilities).[7]

Alibek also provides an account of a vaccine-resistant strain of *F. tularensis* weaponized by himself and colleagues at *ONOPB*. He reports that successful tests of bomblets filled with the agent were subsequently conducted in the summer of 1982 on Vozrozhdenie Island in the Aral Sea. A year later, a dried variant of the weaponized agent was proved to be even more efficient.[8] Pasechnik, a leading *Biopreparat* scientist who defected to the UK, informed Leitenberg and Zilinskas, that in 1982–1983, his facility, the All-Union Scientific-Research Institute of Highly Pure Biopreparations (Leningrad), worked with the Omutninsk Chemical Factory to "*develop a F. tularensis formulation that could be manufactured on a large scale*".[9] This may be a reference to the weaponization of the vaccine-resistant strain developed by Alibekov and his colleagues.

Another project at Vostochnyi, which may be related to the above work, again focused on *F. tularensis*. It concerned the creation of large-scale production of Protein A at the Omutninsk Chemical Factory. A small team is reported to have been despatched from the Institute of Immunology (Lyubuchany) to the plant. The resulting process was expensive and the yields were not great. However, sufficient protein was produced to coat enough *F. tularensis* cells to conduct tests in animal models. The idea was to allow the modified cells to bypass the immunodefence system in humans. This project is described by Leitenberg and Zilinskas as one of the five major accomplishments of the Soviet second-generation BW programme in bacteriology. A cover story, stating that Protein A was to be used in the large-scale manufacture of affinity chromatography and immune-enzyme assay equipment, was created for personnel at the plant.[10]

Accidents appear to have occurred during the military work on *F. tularensis* at Vostochnyi. Alibek, for example, reports that in March 1983 there was a leakage involving the agent. However, owing to a prompt response, the microorganism was confined within *ONOPB*'s labs. Fedorov, meanwhile, reported that in the 1980s a facility at Vostochnyi was responsible for the release of a military strain of tularaemia into the environment. Rodents living in the forest surrounding the plant are reported to have become infected and all attempts to eradicate the disease from local populations failed. The search for the source of contamination at the facility allegedly resulted in the discovery of an insignificant leak in a pipe.[11] This story was categorically denied by the Chief Sanitary Doctor of the Kirov *oblast'* at an international seminar held in October 2001 under the auspices of the Volga-Vyatka Centre of Applied Biotechnology.[12]

In the wake of the collapse of the Soviet Union there was, under the auspices of a newly instituted Trilateral Process, a single opportunity for the West to inspect the facilities at Vostochnyi. A US/UK team co-led by Britain's David Kelly began an inspection of Vostochnyi on 10 January 1994. However, the Russian side insisted that the visit was confined to the Omutninsk Chemical Factory and no opportunity was provided to explore the contiguous facilities belonging to *ONOPB*. At all three production facilities visited by the team in Russia, which included Vostochnyi, they identified large-scale production in hardened facilities, aerosol test chambers, excessive containment levels for current activity and accommodation for weapons-filling lines.[13] Russian personnel at Vostochnyi are reported to have been evasive in response to questions.[14]

The Berdsk Scientific Experimental-Industrial Base (*BNOPB*)

Our knowledge of this part of the *Biopreparat* network is fragmentary and unreliable at best. According to Alibek, the Siberian Branch of the All-Union Scientific-Research Institute of Applied Biochemistry came into being in 1975. It may subsequently have been renamed as the Berdsk Scientific Experimental-Industrial Base (*Berdskaya nauchnaya opytno-promyshlennaya baza* or *BNOPB*). The facility is reported to have been engaged in developing filling and assembling technologies used in the production of biological weapons. Its director at around this time was Colonel Vitalii Kundin. Alibekov was transferred to the site in March 1976 alongside Vladimir Rumyantsev and found hundreds of unopened boxes containing the latest Western equipment. In January 1977, Colonel Kundin is reported to have returned to Berdsk from *Biopreparat*'s HQ in the possession of two ampoules containing freeze dried *Brucella* (the causative agent of brucellosis). In an eight-month period, Alibekov and Rumyantsev are reported to have developed a growth medium which could produce sufficient bacteria to allow weaponization. In 1979, Alibekov was promoted to the position of "*acting lab chief at Berdsk*". He left the facility in 1980 and reports the facility suspended operations in 1981.[15] His account is problematic however since it raises the question as to the eventual fate of *BNOPB*. Was it simply dismantled or mothballed? Or could it have been reprofiled and absorbed by the *NIKTI BAV* institute which was also located in Berdsk (see above)? Unfortunately, the US/UK inspection team co-led by David Kelly was unable to provide any answers to these questions. It, rather, was focused on the Berdsk Chemical Factory in October 1993 and was not permitted access to any contiguous facilities at the site.[16]

The Stepnogorsk Scientific Experimental-Industrial Base (*SNOPB*)

Construction, using convict labour, of the buildings comprising the Kazakh Branch of the All-Union Scientific-Research Institute of Applied Biochemistry (*Kazakhskii filial' VNIIbiokhimmashproekt*) began at a site 14 km north of Stepnogorsk in 1982. It occupied territory immediately adjacent to the *Progress* Chemical Plant, where a warehouse complex had previously been located. Sometime after 1982 it was renamed the Stepnogorsk Scientific Experimental-Industrial Base (*Stepnogorskaya nauchnaya opytno-promyshlennaya baza* or *SNOPB*)—see Fig. 5.1. The facility was referred to as P.O. Box 2076 (*p/ya 2076*) in *Biopreparat*'s coded cables. As detailed above, staff at *VNIIbiokhimmashproekt* are reported to have become under immediate pressure from Belyaev, head of *Glavmikrobioprom*, for resolving technical issues associated with safety of personnel and protection of the environment at the facility. Belyaev is reported to have personally monitored progress on this project, pointing most strongly to the particular importance attached to this new facility by the Soviet leadership.[17] The Russian State Archive of the Economy reveals that, by 1985–1986, *SNOPB* was accounting for just over one-fifth of *Biopreparat*'s entire budget for capital construction projects.[18] One report recently suggested that the Soviet authorities may ultimately have invested a total amount of 1 billion roubles in the project.[19]

Fig. 5.1 View from Building 221 of Bunkers 241-244 (foreground) and Building 600 (behind bunkers) and Building 231 (background), *SNOPB*, Stepnogorsk, Kazakhstan, 14 July 1995 (©A. Rimmington)

Eduard Ivanovich Perov was the first director and managed the construction phase of the project. He was initially joined by a further 15 specialists at the construction site. He was subsequently appointed head of the adjacent *Progress* facility and was replaced as director of *SNOPB* for a short period by Colonel (Reserves) Yurii Petrovich Davydkin. There then arrived one of the pivotal figures in the development of the *SNOPB* facilities, Kanatzhan Baizakovich Alibekov (by this time a Major), who was appointed director of the Stepnogorsk Scientific Experimental-Industrial Base in 1983. He was transferred to *SNOPB* from *Biopreparat*'s *ONOPB* facility which he had headed in Vostochnyi (Kirov *oblast'*).

A year later, Alibekov was joined in Stepnogorsk by Colonel (Reserves) Gennadii Nikolaevich Lepeshkin—nicknamed "Guenna" among friends (see Fig. 5.2).[20] The latter was appointed as head of *SNOPB*'s Biosafety Division. Born in Murmansk *oblast'* in 1946, Lepeshkin had previously been employed within the USSR Ministry of Defence's Scientific-Research Institute of Microbiology in Kirov. Eventually, in late 1987 after the

Fig. 5.2 Colonel (Reserves) Gennadii Nikolaevich Lepeshkin (Director *SNOPB*, 1987–1991), standing second-left, *SNOPB*, Stepnogorsk, Kazakhstan, 14 July 1995 (©A. Rimmington)

departure of Alibekov, Lepeshkin was appointed director at *SNOPB*.[21] A number of interviews with Lepeshkin were conducted by US-based researchers during the period 1996–2000. He is questioned about how he felt at the time working on biological weapons and responded, "*I was offered a job, and I accepted it and went there to work....My feelings about working with biological weapons were positive....I liked the work, I felt that it was very important and promising. And, very necessary....You're learning new things, studying the work of other research scientists, analysing, getting various types of new findings, getting results....I chose the life of a military man, military doctor, microbiologist, a professional military man in the bio technology field, so the work gratified me and I lived out my life...lived a life that pleased me*".[22] However, the researchers also point to the fact that Lepeshkin "*felt clearly uncomfortable when asked why he had not told his family the truth about his weapons work—even now, when he is no longer bound to keep that secret. His grown children still think that he was 'a military doctor.' This comment reveals how he is still trying to partition his former work, life, identity, and representation of self*".[23] The oral history project clearly provides a fascinating insight into the mindset of a leading figure in the *Biopreparat* programme.

The Stepnogorsk facility is reported to have occupied a site some 43 hectares in size and eventually embraced 25 buildings which together formed an integrated technological chain.[24] Strains were initially selected from *SNOPB*'s bacterial culture collection and transferred to a high-containment BSL-3 laboratory on an upper floor of a 27-metre high six-storey structure, the main production facility known as Building 221. Following cultivation in 1000 litre fermenters, the bacterial biomass descended by gravity to feed into 10 large-scale fermenters located on the lower floors, each with 20,000 litres capacity.[25] The bacterial solution from these fermenters was then transferred into seven centrifugal separators, operating at 5000 rpm, which separated the bacterial biomass from the nutrient medium and other waste products. Some 1.5 tonnes of concentrated bacterial slurry could be produced in a single three-day production cycle. Building 221 was also linked by a system of insulated pipes to Building 211 which was able to supply 17 types of nutrient medium and had a reported capacity of 30,000 metric tonnes per annum.[26] Building 221 was managed by Vladimir Alekseevich Mityushkin. Working under him was Viktor Fedorovich Panfilov who was the chief technologist in the fermentation workshop. The latter was awarded the highly prestigious USSR State Prize in 1984.[27]

The biomass produced by Building 221 could be transferred to Building 231 for drying and milling of the resulting cake to generate uniform particle sizes. The dryers were manufactured in the GDR and each weighed over 200 tonnes.[28] The underground Bunkers 241-244 (Fig. 5.3) are reported to have contained filling stations where "*special machines filled the concentrated slurry of pathogenic microorganisms into bomblets and then sealed them. Weaponisation could be completed by installing explosive bursters in the bomblets*".[29] A report by the Swedish Defence Research Agency provides a detailed account of the weaponization process. It notes, "*Loading the bomblets (or the like) with explosive was dangerous work. To avoid a blast that would set off all ammunition, loading the explosive was done individually for each piece of BW ammunition and in a separate room to contain any explosion. In the bunker, there was a small cubicle (each wall 2-2,5 m and high enough for a man to stand straight inside), perhaps a mobile unit, all the inside clad with metal sheeting. Inside it was a work bench (ca 2m long) along one wall of the cubicle and a gas tube (ca 1.5 m tall) standing at the*

Fig. 5.3 View of Bunkers 241-244, *SNOPB*, Stepnogorsk, Kazakhstan, 14 July 1995 (©A. Rimmington)

wall opposite the door. It is possible that this could be the type of 'room' used for loading the explosive charges".[30] The blast-proof walls of the bunkers consisted of reinforced concrete two metres thick—"*capable of surviving a nuclear attack*".[31] The bunkers apparently could also be used to fill bomblets with dry formulations.[32] Biomass could also be directed to Bunkers 251–271 for long-term refrigerated storage. These latter bunkers are reported to have been connected to a railway spur fitted out with an overhead crane and had a small helicopter landing-pad nearby.[33]

Building 600 was *SNOPB*'s research and testing building.[34] Alibek provides this vivid description of the facility: "*...Building 600, was the largest indoor testing facility constructed up to that time in the Soviet Union. It was more than 50 feet* (approximately 15 metres) *high, with two giant stainless steel testing chambers hidden inside. The first chamber, designed to withstand the force of a powerful explosion, would be used to analyse the decay rate and dissemination capacities of aerosol mixtures contained in our germ bombs. The second was for testing animals*".[35] Another description of the chamber for testing BW munitions states that it "*was about 200 cubic metres in*

Fig. 5.4 View of security wall, perimeter of *SNOPB*, Stepnogorsk, Kazakhstan, 14 July 1995 (©A. Rimmington)

volume and had walls 1.6 centimetres thick. It was mounted on an octagonal concrete platform, with a waste trap below. Monkeys, rabbits and other animals kept in cages in the facility's vivarium were used for tests in the aerosol chamber".[36]

Building 277 was a waste treatment facility. Other buildings were concerned with power supply, steam generation, waste treatment and so on (Fig. 5.4).

The Transfer of *B. anthracis* Production from Sverdlovsk to Stepnogorsk

During the 1970s, the USSR Ministry of Defence maintained facilities for the production of *B. anthracis* (the causative agent of anthrax) with a view to its potential use as an offensive agent. On Monday 2 April 1979, an anthrax aerosol was accidentally released from the ministry's Scientific-Research Institute of Bacterial Vaccine Preparations, part of Military Compound Number 19 in Sverdlovsk. This incident led directly to the deaths of at least 68 people in Sverdlovsk itself and to cases of animal anthrax in nearby villages (Rudnii, Bol'shoe Sedelnikovo, Maloe Sedelnikovo, Pervomaiskii, Kashino and Abramovo) to the south-south-east of the city. It was the first major indication in the West that the Soviet Union had embarked upon a major offensive biological weapons effort.[37] The accident led the Soviet authorities to seek to relocate the manufacture of this agent away from major population centres. With this aim in mind, in 1981 Brezhnev is reported to have signed a decree ordering the relocation of the military *B. anthracis* production to *Biopreparat*'s highly remote facilities in Stepnogorsk.[38]

Alibekov headed the anthrax project at *SNOPB*. In 1983, the first of some 65 scientists and technicians from the military's Sverdlovsk facility began to arrive in Stepnogorsk. These included Nikolai Chernyshov, known to colleagues as "Kol'ya", who was an expert in drying methods for *B. anthracis.*[39] Many of the newly arriving personnel were unable to adapt to the harsh climatic conditions on the steppe and later returned to Russia. Nevertheless, by 1984 there were 350 personnel employed at the facility and this figure rose to around 900 people in 1986.[40] Alibek describes his role at *SNOPB* thus: "*Once I'd worked out the technique for its (B. anthracis*—author*) cultivation, concentration and preparation, I was to develop the infrastructures to reproduce it on a massive scale—a goal that had eluded our military scientists for years. This meant assembling batteries of fermenters, drying and milling machines, and centrifuges, as well as the equipment required for preparing and filling hundreds of bombs.…My job at Stepnogorsk was, in effect, to create the world's most efficient assembly line for the production of weaponized anthrax*".[41] Eventually, *SNOPB* is estimated to have incorporated capacity for the production of around 300 metric tonnes of weaponized *B. anthracis* spores per ten-month production cycle.[42]

The management structure at Stepnogorsk was a rather complex one. Both the *Progress* Chemical Factory and *SNOPB* formed part of the *Progress* Production Association. During normal peacetime conditions then, Eduard Ivanovich Perov, director of the chemical factory, would head the association. This meant he had priority acccss to shared key infrastructure such as a power plant and a steam plant and *SNOPB* could therefore only rarely operate at anywhere near full capacity. However, in the event that the Soviet authorities ordered a mobilization of the *SNOPB* facility, Alibekov was to take command of the entire association, effectively "*transforming it into a munitions plant*".[43] He notes that the "*pathogenic weapons that emerged would be poured into bomblets and spray tanks and loaded into trucks for shipment to a railroad station or airfield*".[44]

As a result of the pressure-cooker atmosphere at *SNOPB* and the highly dangerous nature of the pathogens it was working with, Alibek reports that there were, incredibly, one or two accidents *every* week. In one of the most serious incidents, a laboratory technician had been infected with *B. anthracis* through an abrasion in his neck. The cutaneous anthrax which resulted was immediately treated with streptomycin and penicillin antibiotics, but a swelling is reported to have erupted on his chest and then spread across his body. It was only after an abnormally high dose of anthrax antiserum was administered as a desperate measure that the technician recovered from the infection.[45]

Notes

1. Introduction to the *Glavmikrobioprom* archive, *RGAE*, Fond 178, Opis 1, undated document.
2. Surprisingly, the Berdsk factory was struggling most at this moment in time with regard to staff recruitment and had only filled 92 per cent of vacant posts in 1986. Insufficient housing and a lack of suitable nurseries and kindergartens were blamed for problems with staff retention and recruitment. Report of the Main Administration "Biopreparat" on Basic Activities of Industrial Enterprises for 1986 and Explanatory Notes, *RGAE*, Fond 788, Opis' 1, Delo 98, pp. 61–64.
3. Leitenberg and Zilinskas state that, according to Pasechnik, a guarded special BW laboratory was located 400 metres away from the main gate of the Omutninsk Chemical Factory and that it appeared small and unassuming. Leitenberg, Zilinskas, with Kuhn, *The Soviet Biological Weapons Programme*, p. 163.
4. Alibek with Handelman, *Biohazard*, pp. 51–69, p. 300.
5. Ibid., pp. 51–69, p. 300.
6. Mangold, T., Goldberg, J., *Plague Wars*, p. 185.
7. Alibek with Handelman, *Biohazard*, pp. 51–69, p. 80, and p. 300.
8. Ibid., pp. 26–27.
9. Leitenberg, Zilinskas, with Kuhn, *The Soviet Biological Weapons Programme*, pp. 163–164.
10. Ibid., pp. 269–270, 794 (note 38).
11. Fedorov, L., *New Times*, 20 September 1998.
12. Papers presented at the seminar are contained in Trufanov, A.F. (Ed.), *Biotekhnologiya, Ekologiya, Meditsina: Materialy III i IV mezhdunarodnykh nauchnykh seminarov 2001–2002 gg*, International Science & Technology

Centre/State Scientific Institution "Volga-Vyatka Centre of Applied Biotechnology", Express, Moscow-Kirov, 2002.

13. Kelly, D.C., The Trilateral Agreement: Lessons for Biological Weapons Verification, in Findlay, T., Meier, O., *The Verification Yearbook 2002*, VERTIC, London, December 2002, p. 99.
14. Mangold, T., Goldberg, J., *Plague Wars*, p. 198.
15. Alibek with Handelman, *Biohazard*, pp. 59–61.
16. Kelly, The Trilateral Agreement: Lessons for Biological Weapons Verification, p. 99.
17. Mityushev, Zapiski obyknovennogo cheloveka, p. 422.
18. The archives reveal that the bulk of *Biopreparat*'s budget for capital construction projects in 1985/86 was being directed towards Novosibirsk *oblast'*, Moscow *oblast'* and Tselinograd *oblast'* in Kazakhstan. It appears likely that this expenditure was related to construction projects at the Science Production Association *Vektor* (Kol'tsovo), the Scientific Experimental-Industrial Base (Stepnogorsk) and the Science Production Association *Biosintez* (Obolensk), all of which were major *Biopreparat* BW facilities under construction at this time. *RGAE*, Fond 178, Opis 1, Delo 5102 Report of Capital Investment from a Territorial Perspective.
19. Roffey, R., Westerdahl, K.S., *Conversion of former biological weapons facilities in Kazakhstan: A visit to Stepnogorsk*, July 2000, Swedish Defence Research Agency, FOI-R--0082—SE, May 2001, p. 17.
20. Vogel, K.M., Ouagrham-Gormley, S.B., Grunberg, S., Paperno, S., A man named Guenna, *Cultures*, Vol. 1, No. 4, p. 77, http://russian.cornell.edu/bio/Cultures_Vol1_Issue4_page76.pdf, Accessed on 15 October 2020.
21. Korotko ob avtorakh, *Stepnogorsk: Zdes' propisany nashi serdtsa*, p. 202.
22. Vogel, K.M., Ouagrham-Gormley, S.B., Grunberg, S., Paperno, S., A man named Guenna, *Cultures*, pp. 78–80, http://russian.cornell.edu/bio/Cultures_Vol1_Issue4_page76.pdf, Accessed on 15 October 2020.
23. Ibid.
24. Interview with the general director of the National Centre for Biotechnology of the Republic of Kazakhstan, Science Industrial Complex *Biomedpreparat*, Stepnogorsk, 14 July 1995. Nesterenko, S., Kriger, V., Ot supersekretnykh vaktsin k sredstvu bor'by so SPIDom. Takov potentsial Stepnogorskogo kompleksa "Biomedpreparat", *Kazakhstanskaya pravda*, No. 122, 28 May 1993, p. 7.
25. Interview with the General Director of the National Centre for Biotechnology of the Republic of Kazakhstan, Science Industrial Complex *Biomedpreparat*, Stepnogorsk, 14 July 1995.
26. Bozheyeva, G., Kunakbayev, Ye., Yeleukenov, D., *Former Soviet Biological Weapons Facilities in Kazakhstan: Past, Present and Future*, Center for

Nonproliferation Studies Occasional Paper No. 1, Monterey, California, 1999, p. 11.

27. Telezhinskaya, N., *Stepnogorsk. Zavod Progress 70-80-e gody*, Personal blog, natalya-telezhinskaya.ru › 2020/04/05, Accessed on 1 October 2020.
28. Roffey, R., Westerdahl, K.S., *Conversion of former biological weapons facilities in Kazakhstan*, p. 17.
29. Bozheyeva, G., Kunakbayev, Ye., Yeleukenov, D., *Former Soviet Biological Weapons Facilities in Kazakhstan*, p. 15.
30. Roffey, R., Westerdahl, K.S., *Conversion of former biological weapons facilities in Kazakhstan*, p. 16.
31. Ibid.
32. Leitenberg, Zilinskas, with Kuhn, *The Soviet Biological Weapons Programme*, p. 278.
33. Ibid.
34. Interview with the general director of the National Centre for Biotechnology of the Republic of Kazakhstan, Science Industrial Complex *Biomedpreparat*, Stepnogorsk, 14 July 1995.
35. Alibek with Handelman, *Biohazard*, pp. 96–97.
36. Bozheyeva, G., Kunakbayev, Y., Yeleukenov, D. (1999), p. 11.
37. Rimmington, *Stalin's Secret Weapon*, p. 23.
38. Alibek with Handelman, *Biohazard*, p. 82.
39. Ibid., p. 83.
40. Ibid., p. 105.
41. Ibid., pp. 87–88.
42. Leitenberg, Zilinskas, with Kuhn, The Soviet *Biological Weapons Programme*, p. 279.
43. Alibek with Handelman, *Biohazard*, p. 89.
44. Ibid.
45. Ibid., p. 105.

CHAPTER 6

The Creation of *Biopreparat*'s Scientific Base: The R&D Complexes at Obolensk, Kol'tsovo and Leningrad

Soon after its creation, *Biopreparat* acted speedily to create its own scientific base with major R&D institutes being created in Obolensk, Kol'tsovo in Siberia, and in Leningrad. These centres were to conduct cutting-edge research in molecular biology as envisaged in the programmes being drafted by Domaradskii and the other high-level scientists sitting on the *MNTS* (see above).

The first of *Biopreparat*'s research centres to be created was the All-Union Scientific-Research Institute of Applied Microbiology (*Vsesoyuznyi nauchno-issledovatel'skii institut prikladnoi mikrobiologii* or *VNII PM*)—see Fig. 6.1. It was known by the codename P.O. Box No. (*P/ya*) *V-8724* and informally it was referred to as "*Ogarkov's farm*"—after the head of *Biopreparat* and presumably pointing to its location in Russia's rural backwaters. The institute, which focused on bacterial pathogens, was created on 1 January 1974 in accordance with Order No. 67 issued by Belyaev, the head of *Glavmikrobioprom*, on 4 December 1973. On 19 April 1974, the CPSU Central Committee and USSR Council of Ministers issued the corresponding Decree No. 704 "*On Measures for the Acceleration of the Development of Molecular Biology and Molecular Genetics and the Use of their Achievements in the National Economy*".[1] The military nature of *VNII PM* was immediately recognized with the appointment of Major General (Reserves) Dmitrii Vladimirovich Vinogradov-Volzhinskii as the first director. Vinogradov-Volzhinskii was born in 1925 in Leningrad, the

A. Rimmington, *The Soviet Union's Invisible Weapons of Mass Destruction*, https://doi.org/10.1007/978-3-030-82882-0_6

Fig. 6.1 View of Building No. 1, All-Union Scientific-Research Institute of Applied Microbiology (*VNIIPM*), Obolensk, September 1993. (©Stephen J. Bungard, used with author's permission)

son of a military doctor. He graduated from the Military Medical Academy (Leningrad) in 1947 and then worked as a postgraduate within the Academy's Department of Biology and Parasitology. He defended his Candidate's thesis in 1951. He was then based at the USSR Ministry of Defence's Scientific-Research Sanitary Institute in Zagorsk in the 1950s. Later, from 1971 to 1975, he returned to serve as head of the Department of Biology and Parasitology at the Military Medical Academy.[2]

During the period 1974 through to 1976, *VNII PM* was provided with a base in Protvino, located around 16 km to the south-west of Serpukhov in the Moscow *oblast'*. No details are available with regard to the nature of the premises it occupied at this stage, but they are likely to have been rudimentary. The town was the location of the Institute for High Energy Physics which had its own particle accelerator U-70 synchrotron. A site was selected some 11 km or so north-east of Protvino for the construction of the new institute and its residential settlement. It occupied a total area of 620 hectares, 228 hectares of which was utilized for production facilities and 392 hectares for residential housing for the scientists and support staff.[3] During a New Year's Eve party, *VNII PM*'s scientists entered a competition to select a new name for the site. Various names were put in a hat and Vinogradov-Volzhinskii then made the final decision, selecting "Obolensk" after the Obolenskys—who were one of the leading aristocratic Russian families and were reported to originate from this location.[4]

Military personnel were employed on the construction project and began felling timber in 1976. They appear to have been working under the supervision of the USSR Ministry of Medium Machine Building (*Minsredmash*).[5] Both the institute and its nearby residential settlement were constructed at the same time. During the first phase of construction from 1976 to 1983, a temporary laboratory campus (*Vremennyi Laboratornyi Gorodok* or *VLG*) was built consisting of huts in which personnel could work. Hut 7 for example, housed the institute's tularaemia laboratory and was headed by Leonid Abramovich Ryapis. During this period, many of the staff were being bused in from Protvino—in August 1995, the author stayed here overnight in a dilapidated hotel reserved for visitors to Obolensk—and the nearby bioscience township of Pushchino and others caught a shuttle bus from the town of Podol'sk around 64 km from the site. Staff were also based in neighbouring villages. Scientists did not begin to reside in the Obolensk settlement until the mid-1980s.[6] The construction of the Main Laboratory Building No. 1 was also initiated in 1976 and completed in 1986. A vivarium complex, the construction of which began in 1984, was completed in 1988. The Soviet archives reveal that *Minsredmash* had an important role in the construction of production areas at the site.[7]

There was very tight security in place, both at Obolensk and on the approach roads to the complex. A Russian reporter who visited the facility in 1991, for example, reported that Building No. 1 was "*surrounded by a three-tier cordon and a barbed-wire fence. Guards are posted at each fence, at the entrance to the block, and on every floor of the building*".[8] *VNII PM* incorporated its own *GAI* (*Gosudarstvennaya Avtomobil'naya Inspektsiya*—State Automobile Inspectorate) department with subordinate posts located at intervals along the approach roads. As late as June 1993, when the author visited the facility, no road signs were in evidence to direct travellers to the complex at Obolensk. The only reference points which the contact at the facility had provided were *GAI* check-points and several of these were passed before reaching the site. After passing the sign marked "*Obolensk*", there was a final check-point where an individual in uniform on a motorcycle with side-car provided further details to the driver of how to reach the complex. It is located in the midst of a densely forested area.[9]

While Vinogradov-Volzhinskii coped admirably with coordinating the construction of the facility, he lacked any knowledge of molecular genetics. It was therefore of crucial importance to find him a suitable deputy who could manage the scientific side of the programme. This led Belyaev, in 1978, to appoint Domaradskii (see above) on a part-time basis as Vinogradov-Volzhinskii's deputy. Over the next four years, Domaradskii made weekly visits to the institute and all scientific work was his responsibility. It was during this period that he also made a vital contribution to *VNII PM.* For Domaradskii had accumulated an extensive collection of altered bacterial strains sourced both from across the USSR and internationally which were required for genetic research. Kalinin, realizing the pivotal importance of this collection to the R&D programme at Obolensk—it would have been extremely difficult to source the strains from the few, closely guarded, Soviet collections then in existence—ordered Domaradskii to transfer it to *VNII PM.*[10]

It later became apparent that Domaradskii's part-time role could not go on indefinitely and eventually Colonel (Reserves) Vladimir Davidovich Savve was appointed deputy to Vinogradov-Volzhinskii. He was a graduate of the S.M. Kirov Military-Medical Academy (Leningrad), who had formerly been employed within the USSR Ministry of Defence's Sanitary Institute in Zagorsk. However, he soon departed to become Director of a spin-off from *VNII PM*, the Lyubuchany-based Institute of Immunology (see Chap. 8). Savve was replaced as first deputy scientific director by Roman Vladimirovich Borovik, a virologist who had previously been employed at the Kazan Veterinary Institute.[11]

Another important figure to join *VNII PM* in the early stages of its existence was Colonel (Reserves) Konstantin Ivanovich Volkovoi. Born in Evpatoriya (Crimea) on 29 September 1929, he had graduated from the S.M. Kirov Military Medical Academy in 1954. He had then joined one of the USSR Ministry of Defence's BW facilities, eventually heading his own laboratory.[12] From 1973 to 1975 he had served as a deputy director of *Biopreparat* but had then committed some "*serious infringement of the secret regime*".[13] In 1975, he joined *VNII PM* as head of the Department of Genetics. Despite his alleged breach of security at *Biopreparat*, he wielded considerable power. One of his key early roles was with regard to the recruitment of talented personnel, especially among the younger generation of scientists.[14]

In February 1982, probably the single most important scientific appointment to *VNII PM* was implemented, when Rychkov, the head of *Glavmikrobioprom*, ordered Domaradskii (see above) to take up the permanent position of second deputy director of science at the institute. This querulous, sometimes erratic, but highly brilliant scientist was the one individual with the capability to drive forward *Biopreparat's* programmes for the creation of a second generation of biological weapons. Domaradskii took up residence alone—his family was left behind in Moscow—in a "*splendid*" two-room flat in the centre of Protvino where he was granted access to the special grocery store *Ryabinka*. This was usually reserved for use only by elite party figures and the senior management of the Institute of High Energy Physics. The store was stocked with unimagined luxuries such as instant coffee and caviar. In a society where there were widespread food shortages, access to such stores was viewed widely among the general population as an enormous privilege.[15]

The importance of the *VNII PM* project in the *Biopreparat* system is highlighted by the fact that on a near monthly basis, no less a personage than Yurii Kalinin headed a commission which visited the facility. It included representatives from both the *VPK* and the Central Committee of the CPSU, together with leading research scientists from other parts of the network. A cortege of vehicles "*with lights flashing and sirens wailing*" would be met by the head of *VNII PM's* security services on a nearby approach road and escorted to the facility. Domaradskii would present the results of activities at Obolensk and then receive advice and feedback from members of the commission.[16]

In September 1983, Vinogradov-Volzhinskii was dismissed from his position as director at *VNII PM*. There had been a sharp deterioration in relations between him and Rychkov, head of *Glavmikrobioprom*.[17] However, the appointment of military personnel to high-ranking management positions within the institute continued when his replacement was announced as Major General (Reserves) Nikolai Nikolaevich Urakov.[18] The latter had previously been employed for 26 years at the USSR Ministry of Defence's Scientific-Research Institute of Epidemiology and Hygiene (Kirov). He had risen to the position of deputy director of the military biological institute.[19] In his former position, Urakov had become closely familiar with biological weapons development and therefore appeared perfectly suited to his new role at *VNII PM*. However, his highly autocratic and controlling style grated with his scientific colleagues and appears to have severely stymied innovative research at the *Biopreparat* institute.[20]

Urakov strengthened the representation of the military at *VNII PM* and, at the end of 1984, appointed Colonel (Reserves) V.S. Tarumov as deputy scientific director. The latter had previously been employed at the military's BW facility in Sverdlovsk and was a tularaemia specialist. He was joined by another military scientist specializing in this organism, Colonel (Reserves) Gennadii Anisimov, who subsequently was appointed head of *VNII PM*'s tularaemia department.[21] Domaradskii reports how Urakov and his ex-military colleagues were uninterested in fundamental science. In one encounter, for example, Urakov stated, "*I don't need your strains, I just need one strain!*" and in another, "*We are not playing here, we are making a weapon*".[22]

By this time considerable progress had been made on the construction of *VNII PM*'s vast Main Laboratory Building No. 1 which occupied an area of 30,000 m^2.[23] It comprised of nine floors, the first two of which were concerned with administration and control rooms maintaining heating, refrigeration and gas supply. The third through to the eighth floors were focused on pathogens, each floor being restricted primarily to a single bacterial genus (*Bacillus*, *Brucella*, *Burkholderia*, *Legionella*, *Francisella* and *Yersinia*). There was a small fermentation hall on the ninth floor, with fermenters of up to 100-litre capacity, together with some elements of the buildings air-handling system. In total, there were 200 laboratory rooms which received supplies of filtered air and had negative air pressure.[24] However, even as late as 1986, only five of the nine floors were in a state of readiness and even these had still not been adequately prepared for work with highly dangerous pathogens. Nevertheless, despite this state of affairs, work had already been initiated on the third floor—which incorporated high-security containment labs—and immediately two or three cases were recorded of workers being infected with tularaemia.[25]

The fourth floor of Laboratory Building No. 1 also housed both a very large culture collection, the Central Collection of Microorganisms, and the Static Climatic, Horizontal Dynamic Device SC-10-HDD-600. The collection was created in 1977 and included around 2000 strains including recombinant cultures.[26] Meanwhile the SC-10-HDD-600 device was a machine set up in 1987 to allow the study of the effects of aerosolized formulations of bacterial pathogens on animals held within its chamber, while varying parameters such as humidity, air-flow rate, temperature and so on. It incorporated six places for holding large animals and an additional 30 places for small laboratory animals. The device was subject to BSL-3 level of protection.[27]

Accomplishments of *VNII PM* in Soviet Weaponization Programmes: *Bonfire* (*Koster*)

Domaradskii presents detailed evidence in support of his contention that *VNII PM* was one of the most militarized within the *Biopreparat* system. As is detailed above, the senior management of the institute was dominated by personnel who had previously served in military BW facilities. This meant that "*when it came to solving 'special' problems—i.e., the actual development of biological weapons strains—the Institute of Applied Microbiology was well ahead of the other institutes in the Biopreparat system, though the Obolensk expertise in technological developments had been largely imported by Urakov and company from the Ministry of Defence*".[28] By contrast, with regard to achievements in fundamental science in molecular biology and genetics and so on, Domaradskii, who through various appointments had an excellent overview of the entire network, argues that Obolensk "*was the worst—the most poorly run and the least productive—of Biopreparat's facilities*".[29]

Leitenberg and Zilinskas provide a very detailed account of *VNII PM*'s weaponization projects. The importance of the institute is evidenced by the fact that it is reported to have accounted for three out of the five accomplishments of the Soviet second-generation BW programme in bacteriology. In 1986, under the secret programme *Bonfire* (see above), *VNII PM* researchers succeeded in genetically engineering a strain of *Bacillus anthracis* which was resistant to seven or eight antibiotics. Multi-resistant strains of *F. tularensis* (which causes tularaemia), *B. mallei* (which causes glanders) and *B. pseudomallei* (which causes melioidosis) were also created.[30] Another significant achievement of *Bonfire* concerned the insertion of virulence genes from *Bacillus cereus* into *B. anthracis*. This work was apparently completed in 1993 and raised fears that the West's existing defence anthrax vaccines would be ineffective against the new genetically engineered pathogen.[31]

The *Factor* (*Faktor*) Programme at Obolensk

Around three years after the commencement of *Bonfire*, a substantive new offensive BW programme named "*Factor*" ("*Faktor*") is reported to have been launched in the Soviet Union which initially aimed to create genetically engineered bacterial and viral agents with the ability to produce various peptides (in particular, neuropeptides) and toxins, which would yield strains with wholly new and unexpected pathogenic properties. In this way classical BW agents could be made even more virulent. The Soviet military apparently considered that these new agents would be essentially undetectable, and might be able to circumvent the provisions of the 1972 Biological and Toxin Weapons Convention.[32] Some senior figures even believed that such genetically engineered agents could be used to disable human targets in much the same manner as teargas and related agents The *Factor* programme appears to have been closely related to, and run in parallel with *Bonfire*.

The military scientist credited with the formulation of the *Factor* programme is Major General Igor' Petrovich Ashmarin, who was born on 20 September 1925 in Leningrad and was a graduate of the city's S.M. Kirov Military-Medical Academy. Ashmarin later served as a deputy director at the Ministry of Defence facility in Zagorsk. Here, at a scientific conference held at the beginning of the 1960s, he had supported papers on the genetic code which came under attack from supporters of Lysenko. Later, he participated in a country-wide campaign against Lysenko's theories. During the period 1976–1986, he served as a deputy of Efim Ivanovich Smirnov in the USSR Ministry of Defence's Fifteenth Administration—which was responsible for the military's BW programme.

Ashmarin pioneered research in the Soviet Union on a new class of bioregulators, neuropeptides. He was responsible for the launch of the Interdepartmental Programme "Neuropeptides" ("*Neiropeptid*") which he headed from 1977 to 1980. Later, at a joint session of the general meeting of the USSR Academy of Sciences and the USSR Academy of Medical Sciences on 19–20 November 1980, Ashmarin reported on Soviet programmes in small peptide research.[33] He also revealed details of an ambitious new initiative, noting that the All-Union "Neuropeptide" ("*Neiropeptid*") Programme of the USSR Academy of Sciences and the All-Union Programme "Physiologically Active Peptides" ("*Fiziologicheski aktivnye peptidy*") of the USSR Academy of Medical Sciences, were being amalgamated. The new combined effort he announced would embrace the work of more than 20 facilities within the USSR. He emphasized in his address that the programme of research in this area must be expanded and that coordination must be improved. Ashmarin also called for the "*creation of a broad base for the experimental-industrial synthesis of new peptide preparations*".[34]

Ashmarin interacted closely with Domaradskii and together they appear to have coordinated *Biopreparat*'s research activities within the *Factor* programme. One of the major successes attributed to *VNII PM*, undertaken as part of the military project, concerned the genetic engineering of *Legionella pneumophila* (the causative agent of Legionnaire's disease). This work was led by Sergei Gennadievich Popov who arrived at Obolensk in 1986 and was appointed head of the Department of Genetically Engineered Preparations. Popov began work on the production of synthetic genes using chemical synthesis. From 1986 to 1987, Popov and his team succeeded in developing a synthetic gene that coded for the production of a myelin peptide and to insert this gene into an expression plasmid. In an interview conducted in September 2003, he reported, "*The hope was, making those constructs, it would be possible to provide bacterial agents and viruses completely new properties which they did not have in natural conditions. So, for example, a virus could produce something absolutely difficult to imagine in natural circumstances, like peptides which destroy the immune system in a very special way*".[35]

Eventually, *L. pneumophila* was selected for introduction of the plasmid. Popov's goal was then to use this genetically engineered bacterium to produce a protein resembling myelin during an infection. The human immune system would then not recognize myelin as a normal component and target it for destruction as a foreign body. In the inflammatory response which would then be triggered, the myelin surrounding nerve cells would be attacked along with brain cells which make and repair the myelin. The result would be an "*artificial version of multiple sclerosis*".[36] The modified strain carrying the myelin genes was subsequently injected into guinea pigs which at first exhibited symptoms of mild pneumonia. Later the guinea pigs suffered brain damage and paralysis, then died with a mortality rate approaching 100 per cent. When Popov observed what had happened to the animals during testing, he was apparently "*overwhelmed with doubt*".[37] Nevertheless, he and his colleagues ushered Urakov over to the vivarium to witness their grim success. Sometime around 1989, in Obolensk, Popov presented his findings to a group of more than 50 *Biopreparat* leaders, senior scientists and military officials. Apparently, the Soviet military responsible for overseeing the *Biopreparat* offensive programme were less than enthusiastic about this futuristic research and the recombinant *Legionella* was never made into munitions or deployed in any way.[38] The US and its allies were subsequently provided with an opportunity to learn much more about Popov's research when, in 1992, he left Russia for a visit to the UK and never returned. Initially studying molecular biology in Cambridge, he was subsequently employed in the US.

Biopreparat's role within the *Factor* programme was not confined to Obolensk and other facilities under its control, such as the Institute of Highly Pure Biopreparations (Leningrad) and the All-Union Scientific-Research Institute of Molecular Biology—*VNII MB*—in Kol'tsovo (see below), were assigned specific tasks. In 1985 for example, Ernst Georgievich Malygin, a head of department at *VNII MB* was awarded the Lenin Prize, among the highest awards in the Soviet Union, for his work on the use of genetic engineering methods to create a microbial strain producing biologically active peptides.[39] Malygin's background is as a chemist having graduated in 1960 from the D.I. Mendeleev Moscow Institute of Chemical Technology. He then completed his Candidate's degree at the USSR Academy of Sciences Siberian Department's Novosibirsk Institute of Organic Chemistry where he remained until 1975 before transferring to Kol'tsovo.[40] It should be noted that the Defence Department of the Central Committee of the Communist Party of the Soviet Union played a pivotal role in drafting the lists of those to be awarded key honours such as the Lenin Prize.[41] As a consequence leading defence scientists working within the Soviet military-industrial complex on nuclear, chemical and biological programmes were showered with prizes and honours by the Soviet authorities.

Mention should also be made of the fact that late in the Soviet period, *Biopreparat* was also organizing a facility for the production of neuropeptides. In the absence of any concrete information one can only speculate, but this may very well have been an attempt to commercialize some of its ongoing R&D in this area. Whatever the case, the archives reveal that on 13 April 1990, the USSR Ministry of the Medical Industry issued an Order regarding the organization of a science production department for the development and production of new biological preparations named "hypnoproteins". The department was to be based at the State Concern *Biopreparat*'s Kharkov Enterprise for the Production of Bacterial Preparations and consisted of a scientific-research laboratory and an experimental-technological section for production. The production section was to be based at both the Kharkov Enterprise for the Production of Bacterial Preparations and the Science Production Association *Ukrmedbioprom*'s Kharkov Chemical-Pharmaceutical Factory. Evgenii Petrovich Naimytenko was appointed as head of the new project. He was a Candidate of Veterinary Sciences and head of Department at the Kharkov

Zooveterinary Institute. The two industrial factories were to be provided with the necessary financial resources to establish new production areas and biochemical and bacteriological laboratories and related equipment. Kalinin, chairman of the board of *Biopreparat* at this time, was given specific responsibility for the execution of the Ministry's order relating to this project.[42]

The 1990 order appears to be highly significant and appears to be one of the first times that a Soviet-era industrial factory has been linked to the research and production of neuropeptides. The director of the Kharkov Enterprise for the Production of Bacterial Preparations, G.A. Sennikov, and the director of the Kharkov Chemical-Pharmaceutical Factory, V.N. Fedorov, were ordered to bring forward a proposal for the creation on the basis of their facilities, of an Inter-Branch Scientific-Technical Complex for the Creation and Production of New Biological Preparations Based on Phospholipids and Hypnoproteins. Such complexes were extremely rare in the Soviet Union and usually created by the authorities only for the pursuit of high-priority projects. Later, on 21 November 1990 the USSR Ministry of the Medical Industry issued Order No. 388 regarding the creation under its auspices of the Small State Enterprise "Neuropeptide" (known by the Russian abbreviation of *MGP "Neiropeptid"*). The enterprise was formed on the basis of a Science Production Department for the Development and Production of New Biological Preparations—Hypnoproteins—itself based at the Kharkov Enterprise for the Production of Bacterial Preparations. The preparations being developed at Kharkov are reported to have possessed immunostimulating, analgesic and psychotropic effects. Evgenii Petrovich Naimytenko (see above) was appointed director of the small enterprise.[43] Whatever the reason for the creation of *MGP Neiropeptid*, it did not remain for long under *Biopreparat* control and on 10 April 1991 it was transferred to another newly created bioscience conglomerate called the All-Union Association *"Farmbiopress"*.[44]

The Transfer of the Lotoshino Production Plant to *VNII PM*

Sometime around 1987, *VNII PM* was selected as the lead institute in the newly created Science Production Association *Biosintez* (*NPO Biosintez*). Besides *VNII PM*, *Biosintez* also incorporated the Institute of Immunology

(Lyubuchany) and a major industrial plant, the Lotoshino Biochemical Factory (*Lotoshinskii biokhimicheskii zavod*) located on the river Lob', 35 km from Volokolamsk in Moscow *oblast'*. The idea behind the creation of the new associations in the USSR was to promote the transfer of novel technologies and processes from R&D institutes to industrial production units. As part of the reorganization, on 8 June 1990 the USSR Ministry of the Medical Industry issued Order No. 212, renaming the Lotoshino Biochemical Factory as the Lotoshino Experimental Biotechnology Enterprise (*Lotoshinskoe opytnoe biotekhnologicheskie predpriyatie* or *LOBP*).[45]

LOBP in fact had its origins in August 1966 when the Lotoshino Ethanol Factory (*Lotoshinskii spirtovoi zavod*) was renamed as the Lotoshino Biochemical Factory and transferred to the Main Administration of the Microbiological Industry under the USSR Council of Ministers (*Glavmikrobioprom*).[46] Ostensibly, in its new role affiliated to *Biopreparat*'s *VNIIPM* it was intended to be used by the Obolensk institute for the production of a range of new preparations, including drugs using traditional and genetic engineering technology, immunobiological diagnostics, nutrient media bases, plant protection agents and veterinary preparations. It was only more than a decade later, in February 2002, that some additional information, relating to BW activities at the facility was revealed. The Russian daily newspaper *Moskovskii komsomolets* reported that at some point during the Soviet period, the Lotoshino facility had produced some strains of anthrax. These are then reported to have been transported in special containers to the USSR Ministry of Defence's BW proving ground on Vozrozhdenie Island in the Aral Sea. Here military personnel are reported to have combined them with "*assembly components*" from other plants and made a final product.[47] Whether *Biopreparat* or indeed *VNII PM* had any role in the alleged BW activity at the plant is unknown.

Biopreparat's Virology Centre: The All-Union Institute of Molecular Biology (*VNII MB*)

Just a few months after the establishment of its bacteriology research centre in Obolensk, *Biopreparat* acted to create a new dedicated virology R&D facility in Kol'tsovo (see Fig. 6.2), located about 20 kilometres south-east of Novosibirsk. On 2 August 1974, *Glavmikrobioprom* issued Order No. 1683 to create the All-Union Scientific-Research Institute of Molecular Biology (*Vsesoyuznyi nauchno-issledovatel'skii institut molekulyarnoi biologii* or *VNII MB*).[48] The facility was also known by the

Fig. 6.2 Academician Sandakhchiev Avenue, Kol'tsovo, Novosibirsk *oblast'*, 2018 (K. Artem.1, Sandakhichiev Prospekt, Koltsovo, Novosibirsk Oblast, https://commons.wikimedia.org/wiki/File:Sandakhchiev_Prospekt,_Koltsovo_01.jpg, This file is licensed under the Creative Commons Attribution-Share Alike 4.0 International license, Accessed on the 28 September 2020)

codename of P.O. Box *V-8036* (*p/ya V-8036*). A commission headed by the chairman of the Novosibirsk Regional Executive Committee (*oblispolkom*) was established to consider various sites for the location of the institute, including Iskitim, the very isolated and remote settlement of Tashara, and Berdsk (see above).[49] Factors that favoured the eventual selection of Kol'tsovo included the fact that it was downwind of the major population centre of Novosibirsk and in the event of any accidental release of viruses they would be carried safely away from the city. A railway station was in close proximity at Krakhal' and it was also located just 10 km from the major science centre of *Akademgorodok*, home, at its peak, to 65,000 scientists and their families. The proximity of such a powerful scientific centre, which included a substantive bioscience element, was a major factor in the selection of the Kol'tsovo site.[50]

During the first four years of its existence, *VNII MB* was provided with temporary accommodation in *Akademgorodok* (see Fig. 6.3). Its laboratories were based within a number of local facilities belonging to the Siberian

Fig. 6.3 View of *Akademgorodok* (Novosibirsk)—the Siberian scientific centre, 30 April 2005 (Elya, Picture of Novosibirsk Akademgorodok—the Siberian scientific centre, 30 April 2005, https://de.wikipedia.org/wiki/Bild:Luftbild_akademgorodok.jpg, Permission is granted to copy, distribute and/or modify this document under the terms of the GNU Free Documentation License, Version 1.2 or any later version published by the Free Software Foundation; with no Invariant Sections, no Front-Cover Texts, and no Back-Cover Texts. A copy of the license is included in the section entitled GNU Free Documentation License, Accessed on the 6 October 2010)

Branch of the USSR Academy of Sciences (*Sibirskoe otdelenie Akademii Nauk SSSR* or *SO AN SSSR*) including the Novosibirsk Institute of Bioorganic Chemistry, the Institute of Organic Chemistry, the Institute of Catalysis and the Institute of Cytology and Genetics. The initial interactions between *VNII MB* and these institutions subsequently formed the basis of long-term collaborative programmes.[51] The majority of the institute's staff were initially recruited from the academy's Siberian branch, many of these were graduates of Novosibirsk State University. They are reported to have been subjected to special security checking prior to beginning work at the new facility and to have been instructed not to discuss their work with outsiders. The military nature of the institute is reported to have quickly become apparent to civilian academy-based scientists.[52] *SO AN SSSR* also transferred one of its buildings located on Detskii proezd, 8 in *Akademgorodok* to *VNII MB*, which became its temporary headquarters.[53]

The first acting director of *VNII MB* was Nikolai Ivanovich Partrikeev, an experienced builder who had previously headed the Construction Administration of the Novosibirsk *sovnarkhoz* (see above). Besides having the necessary experience with regard to the commissioning of large-scale projects, he also had high-level party contacts in the Novosibirsk region, which were necessary for such an ambitious undertaking.[54] He drew on convict labour for the construction of *VNII MB*'s research and ancillary buildings which were located in a secret new settlement. The design of the new facility's research and production units was completed by *Biopreparat*'s *VNIIbiokhimmashproekt* (see above). A large military and civilian construction enterprise, *Sibakademstroi*, was charged with the construction of *VNII MB*. Such was the scale of the project that it created a special subordinate unit, the Construction Administration No. 10 (*stroitel'noe upravlenie No. 10* or *SMU 10*) for this specific task. The first building at the site was commissioned in April 1978 and incorporated laboratory areas. On 30 August 1979, the Novosibirsk Regional Executive Committee (*oblispolkom*) officially named the site "Kol'tsovo" after the internationally renowned Russian biologist and a pioneer of modern genetics, Nikolai Konstantinovich Kol'tsov. Later, in 1982, Building 1, which was *VNII MB*'s main research centre and included BSL-3 and BSL-4 high containment laboratories on five floors, began operation at the site. A shuttle bus linked *VNII MB*'s temporary accommodation at Detskii proezd, 8, *Akademgorodok* to Kol'tsovo.[55] Construction of accommodation for the staff also began at the institute's nearby residential area located close to Baryshevo on the Inya river.[56] By 1979, 1700 staff and their families were living in accommodation built at the site.[57]

The intimate links between the Kol'tsovo institute and *SO AN SSSR* were crystallized with the appointment in 1979 of Lev Stepanovich Sandakhchiev as director of *VNII MB*. Sandakhchiev, who was born on 11 January 1937, in Rostov on Don, had later read technical chemistry at the D.I. Mendeleev Moscow Institute of Chemical Technology. For the next decade and a half, he was an employee at *SO AN SSSR*'s Institute of Organic Chemistry (Novosibirsk). By 1970 he was heading a laboratory at the institute.[58] Sandakhchiev must have been aware of the new opportunities which would be made available to conduct leading edge research at the newly created *VNII MB* and in 1974, he joined the facility as a head of department.[59] Later, from 1975 to 1979 he served as assistant scientific director under Partrikeev. He then served as director of the institute from 1979 to September 2005.[60] Sandakhchiev died on 29 June 2006.[61]

VNII MB appears to have had fewer staff recruited from military biological facilities than was the case in Obolensk. However, the numbers were still significant and *Biopreparat* ensured that there was military representation at a high level of management. This is evidenced by the fact that in 1979, Nikolai N. Sergeev was appointed to the critical post of first deputy scientific director at the new institute.[62] He is known to have previously headed a scientific group at the USSR Ministry of Defence's dedicated virology facility, the Scientific-Research Sanitary Institute in Zagorsk.[63] Here he had focused on Variola major (the causative agent of smallpox) and haemorrhagic fever viruses. Frustratingly, there is remarkably little trace in open publications of the period he spent at the Kol'tsovo facility. The same is true of Sergeev's replacement, Nikolai B. Chernyi. All that is known of him is that he too was from the military and also originally based at the Zagorsk Sanitary Institute. Later he published a number of papers on the Marburg virus.

The Creation of the Science Production Association *Vektor*

In 1985, in parallel with changes taking place at other *Biopreparat* R&D facilities, a major reorganization was undertaken with regard to the management of activities at Kol'tsovo. On 25 March, *Glavmikrobioprom* issued Order No. 20 regarding the creation of the Science Production Association *Vektor* (*NPO Vektor*). Sandakhchiev was appointed General Director of the new association and his institute, *VNII MB*, was made its lead facility.[64] This restructuring coincided with the launch of an ambitious Five-Year Plan issued by CPSU General Secretary Mikhail Gorbachev, which outlined a major acceleration of BW programmes in the Soviet Union and provided for the allocation of US$1 billion equivalent through to the end of the decade.[65] It was against this background that a major expansion of BW activities began in the Novosibirsk region. The first manifestation of this was the issue on 27 November 1987 by *Glavmikrobioprom* of an order for the incorporation within *Vektor* of the Scientific-Research Design Institute of Biologically Active Substances (*Nauchno-issledovatel'skii konstruktorsko-tekhnologicheskii institut biologicheski aktivnykh veshchestv* or *NIKTI BAV*).[66]

With the new funding from Gorbachev, *Biopreparat* also began a rapid expansion of *Vektor*'s animal breeding and manufacturing capabilities. In

1984, *Vektor* was given control of the state-owned farm Iskitimsky (Morosovo, near Iskitim). This embraced 5035 hectares of land and both housing and existing production buildings. The latter were subsequently used as the basis for the construction, in September 1989, of *Vektor*'s new Experimental Agricultural Production Enterprise (EAPE) located in Morosovo. In 1988, the entire village of Nizhny Koyon with an adjacent land area of 7511 hectares was transferred to *Vektor*. The EAPE was used for the breeding of large laboratory animals including cattle, horses, sheep, donkeys and pigs which were used for both research and production of blood sera. Some wild animals were also bred here, including marmots, deer and badgers. The facility was also used for milk and meat production for members of *Vektor*'s staff. Apartments for personnel were also built on the newly acquired land along with a middle school for their children.[67] On 31 March 1988, *Vektor* also issued Order No. 34 for the creation in Kol'tsovo of a new Scientific Experimental-Industrial Base (*Nauchnaya opytno-promyshlennaya baza* or *NOPB*).

The total area allocated for construction of *VNII MB* and *NOPB* buildings was 139 hectares. During the first phase of construction from 1975 to 1991, 211 million roubles were expended (1984 prices). Research and experimental buildings with a total area of 82,000 m^2 were commissioned during the same period which included: Laboratory Building No. 1 (25,700 m^2), Laboratory Building No. 5 (4900 m^2), Laboratory Building No. 6 (6600 m^2), Laboratory Building No. 6a (6600 m^2), Experimental Building No. 13 (17,000 m^2), Experimental-Technological Building No. 15 (7600 m^2), Nursery of Laboratory Animals No. 104 (6800 m^2) and Nursery of Laboratory Animals No. 104/1 (6800 m^2). A second construction phase was initiated in 1986 which focused on the enlargement of the research and production base.[68] Other buildings constructed at the Kol'tsovo site included Buildings 19 and 20 (waste disposal), Building 10 (incinerator), Building 9 (laundry), Building 38 (canteen) and Building 17 (guardroom).[69]

Building 6—the Biocontainment Virological Laboratory Experimental Building, and its twin, 6a, were the location of BSL-4 high containment laboratories. These were designed for experimental investigation of specific viral pathogens. Each building incorporated total floor space of 6336 m^2, including:—working floor space of zone 3 (biocontainment level BSL-4) 1440 m^2, working floor space of zone 2 (biocontainment level BSL-3) 720 m^2, working floor space of the clean zone laboratories 720 m^2, and working floor space of the technical premises for maintenance and biocontainment systems 2448 m^2. Each building was equipped with ventilation systems which maintained negative pressure, hermetic entrances, double cascade filters for fine purification, a system of waste piping capable of collecting up to 32 m^3 of waste water and of treating it at a temperature of 135°C and with a productivity rate of 16 m^3/hour. The transfer units (autoclaves AGP1200/2,5 and disinfecting chambers KDF-4) for the treatment of waste, suits and instruments operated at a temperature of 135 °C. The buildings also incorporated systems dispensing disinfecting solutions (6 per cent hydrogen peroxide and 4 per cent sodium hydroxide) for treatment of the entire working area, with isolating boxes (biocontainment level P-4) and three independent power supply sources.[70]

Buildings 6 and 6a incorporated unique animal testing equipment with the rather awkward name of the Climatic Static-Dynamic Unit UKSD-25. These units had been brought into operation in 1986. Each was supplied with "*devices for dispersion of the preparations under study, their maintenance in the aerodispersed state, and with precision batch measuring boxes which provides the high accuracy of delivery of the required dose of aerosol to laboratory animals. Both small (for example, mice for which there are 6 boarding of 20 seats) and big (rabbits, monkeys for which there are 6 boarding) laboratory animals can be used in the unit*".[71] All the operations of the units could be performed at the BSL-4 level of protection. The unit was equipped with a ventilation system, systems of waste piping and disinfection which provided deactivation of the waste, and with three independent power supply sources.[72] According to Leitenberg and Zilinskas, these units were "*strong enough to withstand the explosions needed to disperse viral formulations….The final testing stage required the use of nonhuman primates (including apes)*".[73]

Another unique capability belonging to *Vektor* was a laboratory animal facility, dubbed the *Laboratory Animals Farm*, housed in a five-storeyed building with a total floor space of 4225 m^2. Opened in 1986, it was designed for the production and maintenance of pure lines of laboratory animals which were used in research experiments. Each floor of the building contained 10 box stalls for animals with 14-fold air exchange. The facility comprised a feed-preparation unit housed in a two-storey building with a total floor space of 1700 m^2; a storehouse for animal foodstuffs with a total floor space of 1200 m^2; and a one-storey building for the care of rabbits, goats, sheep and fowl with a total floor space of 600 m^2. This was one of only two facilities in the USSR—the other being located in

Sukhumi, Georgia—which bred monkeys for medical research. The technicians at the Kol'tsovo facility succeeded in breeding several generations of green monkeys.[74] *Biopreparat*'s increasingly important role in the breeding of laboratory animals was recognized in May 1990, when Alibekov was appointed by the USSR Ministry of the Medical Industry as chair of an organizing committee for a meeting focused on their supply and use (see below).[75]

Some sense of the scale of activities at *Vektor* is provided by the numbers of personnel employed. By 1990, the association as a whole is reported to have had some 3924 personnel on its staff. Of these, by far the largest part, 2617, were employed within *VNII MB*. Out of the total figure, a further 646 staff were based at *NOPB* and 388 at *NIKTI BAV* (Berdsk). The EAPE (Morosovo) accounted for the final 273 personnel.[76]

Accomplishments of *VNII MB* in Soviet Weaponization Programmes

The initial focus at Kol'tsovo was on Marburg virus and Ebola virus, with both pathogens perceived as having enormous potential for use in weaponization programmes. As was the case at Obolensk, one of the first problems facing *Vektor* was to source the viruses required for its BW R&D. The military BW facility at Zagorsk, which focused on viruses, might be expected to assist the *Biopreparat* institute. However, the USSR Ministry of Defence, which had assembled its culture collections over more than two decades, was fiercely protective of its holdings. Despite this, Leitenberg and Zilinskas suggest that the Zagorsk facility did eventually transfer a Marburg variant to Kol'tsovo. *Vektor* also turned to the Belorussian Scientific-Research Institute of Epidemiology and Microbiology (*Belorusskii nauchno-issledovatel'sku institut epidemiologii i mikrobiologii* or *BelNIIEM*)—which operated under the Belorussian SSR Ministry of Health—for assistance in this regard. Under *BelNIIEM*'s Director Veniamin Iosifovich Votyakov, the Minsk institute had organized a Department of Virology which initially incorporated the following laboratories: Influenza and Other Respiratory Infections; Poliomyelitis; Enteroviral Infections; Tick-Borne Encephalitis; Viral Hepatitis and others.[77] Later, in the early to mid-1980s, *BelNIIEM* established close collaborative links with the Institute of Tropical Medicine (Instituut voor Tropische Geneeskunde)—ITM, in Antwerp which involved exchanges of both scientists and viral strains.[78] There were several joint publications at this time whose authors included prestigious scientists such as Guido van der Groen and Igor' Sergeevich Lukashevich.[79] As part of this collaboration, the Belgian institute, obviously unaware of the Soviet BW programme, provided both Marburg virus (the Voege isolate or MARV-Voe) and Ebola virus (the Mayinga isolate or ZEBOV-May) to researchers at *BelNIIEM*.[80] The variants of the viruses were subsequently transferred to the Kol'tsovo facility.[81]

A key requirement for the successful weaponization of viruses at *VNII MB* was the development of dry viral formulations which could be dispersed in a respirable aerosol. This key work was apparently led by Anatolii Petrovich Sadovskii.[82] Born in 1937, he had graduated from Rostov-on-Don State University and then joined *SO AN SSSR*'s Institute of Inorganic Chemistry (Novosibirsk). Sadovskii was appointed to *VNII MB* in 1978, where he was employed concomitantly as a head of department and head of laboratory before becoming a deputy director at the institute. Officially his work was focused on the stabilization of biological preparations and with the inactivation of biological agents, and on optimization of the application of microbial pesticides in aerosols.[83]

R&D work on the Marburg virus is reported to have begun at the end of 1983 on the fifth floor of *VNII MB*'s Building 1. Initial efforts focused on the creation of a killed virus vaccine but all attempts to create such a preparation are reported to have failed. *VNII MB* researchers were therefore faced with the awful realization that any lapses in safety practices which resulted in infection would almost certainly result in a horrific death. During the first phase of work with the Marburg virus, *VNII MB* researchers discovered that guinea pigs were the best laboratory animal for large-scale propagation of the virus. A second phase of work then followed which succeeded in propagating Marburg virus in high titre in cell culture. Eventually, *VNII MB* scientists are reported to have developed technology for the production of a dry formulation of the virus which could be dispersed in an aerosol.[84]

VNII MB also played an important role in the wider *Biopreparat* programme named *Factor* which initially aimed to create genetically engineered bacterial and viral agents with the ability to produce various peptides (in particular neuropeptides) and toxins, which would yield strains with wholly new and unexpected pathogenic properties. The institute apparently used peptides synthesized both in Kol'tsovo and at other parts of the network, as templates for synthesizing genes that would code for their production.[85] A key scientist at *VNII MB* focused on peptides was Vladimir Vasileevich Samukov. Born in 1953 in Ust-Kamenogorsk (Kazakhstan), he had had subsequently attended the Novosibirsk Physical and Mathematical School. In 1975, he graduated in organic chemistry from Novosibirsk State University and then secured a postgraduate research position at the newly formed *VNII MB*. In 1983, Samukov founded and became the head of the institute's Laboratory of Peptide Chemistry which was engaged in the synthesis of opioid peptides, Diazepam binding inhibitor, LHRH analogues and so on. Eventually, in 1991, Samukov spun-off the *Vektor-BioProdukt* company with the aim of identifying commercial applications of his team's R&D in peptide chemistry.[86] *VNII MB*'s reported involvement in additional weaponization and research programmes, including those relating to the Ebola virus, the Venezuelan Equine Encephalitis virus, Variola virus, Machupo virus and others is documented in very great detail in Leitenberg and Zilinskas, and in the absence of any other sources the author will not attempt to summarize their findings here.

Accidents During *VNII MB*'s Pursuit of Weaponization Programmes

The pursuit by *VNII MB* of the Marburg virus weaponization project led to at least one documented fatality among the institute's staff. Nikolai Vasil'evich Ustinov—known to his close friends as Mikolka—was born on 17 July 1943. His father, Vasilii Sergeevich, had served as a frontline soldier and was an employee of the Tomsk Regional Executive Committee. His mother was a proof-reader with the publishing house of the local polytechnical institute. During the 1950s as a child, Ustinov contracted encephalitis (TBE) after being bitten by an infected tick and was hospitalized for a whole summer. Ustinov then enrolled in the Tomsk Medical Institute where he specialized in paediatrics. During his three years of medical practice in the Tomsk region, he met his future wife, Evgenia. Ustinov though had wider ambitions to make a real impact on healthcare and therefore decided to focus on medical research. He initially joined the Tomsk Bacteriological Institute—later to be renamed as the Scientific-Research Institute of Vaccines and Sera, and forming part of the Science Production Association *Virion* (*NPO Virion*). Here he worked for the next seven years as a virologist and completed his Candidate's thesis on aspects of the treatment of TBE.[87]

In 1983, Ustinov received a call from Vladimir Pekhen'ko, the deputy director of *NPO Virion*. He offered Ustinov a position at the newly created *VNII MB* in Kol'tsovo. Ustinov was one of several scientists the *Biopreparat* institute recruited from Tomsk at this time including Yurii Rassadkin and Boris Chernyi.[88] Ustinov joined *VNII MB* in 1983, at the same time as Aleksandr Alekseevich Cherpunov, and both scientists were immediately despatched for three-month training to the Irkutsk Scientific-Research Anti-Plague Institute of Siberia and the Far East. Ustinov was then appointed head of *VNII MB*'s department of virus production. Ustinov and other colleagues working on dangerous viruses were required to be resident in Kol'tsovo—both he and Cherpunov were allocated apartments in a newly commissioned building there—and could not travel beyond the settlement without first spending two days in quarantine.[89]

In mid-April 1988, Ustinov is reported by Cherpunov to have performed a routine but complex procedure in conjunction with an inexperienced colleague. It involved the extraction of blood from a guinea pig which had been previously injected with Marburg virus. The needle of the syringe had to be inserted precisely into the heart of the animal. Ustinov is reported to have assisted his co-worker by placing his hand under the animal. The needle apparently passed through the animal and into Ustinov's finger. After the accident, Ustinov was immediately admitted into *VNII MB*'s high-containment isolation ward. Gamma globulin was administered although none was available immediately at the institute and it was not until the 20 April that Ustinov received the preparation. The first symptoms presented themselves four days after infection. Ustinov's temperature rose sharply, and he experienced muscle pains, vomiting, diarrhoea and coagulation disorders with bleeding developed.[90]

Ustinov died on 30 April in terrible agony, bleeding profusely and unable to move or even open his eyes. During the period through to his death, Ustinov maintained a log in which he recorded his feelings and symptoms and which he hoped would lead to an improved understanding of the virus and its effects. His corpse was carefully interred with Alibek reporting that it was "*covered with chloramine disinfectant and wrapped in plastic sheeting. The remains were placed inside a metal box, welded shut, and fitted into a wooden coffin*".[91] Russian sources state the coffin was buried to a depth of five metres and covered with concrete. Lev Sandakhchiev, director of *VNII MB*, gave a short eulogy at the graveside. Scientists at

VNII MB subsequently isolated a new Marburg virus variant, named Variant U, from samples of Ustinov's organs and this was absorbed into the institute's culture collection.[92]

The death of Ustinov does not appear to have caused the authorities to curtail in any way the Marburg virus project at Kol'tsovo. Another accident occurred just two years after Ustinov's death, again resulting from *VNII MB*'s research on the Marburg virus. On 11 April 1990, Sergei Vizunov, a junior researcher at the facility working on the virus, presented with symptoms, including severe conjunctivitis, he attributed to a cold.[93] At this point he took some over-the-counter medications to address these symptoms. Under the institute's safety regime, samples were taken from his blood and nasopharynx and injected into guinea pigs. After six days, the animals began to get sick and die. Vizunov's temperature subsequently rose to 38 °C. *VNII MB* researchers believed the slow and uncharacteristic spread of infection in Vizunov resulted from an unusual route of infection through the mucous membrane in the eye. Believing the blood serum of a guinea pig he was handling to have lost its infectivity, Vizunov may have inadvertently touched his eyes with his finger.

Vizunov was immediately admitted to *VNII MB*'s isolation ward. The laborious process of contact-tracing was now initiated. It was determined that prior to infection he had had contact with 12 relatives—presumably including his wife and daughter—and 58 employees based at *Vektor*. The relatives were released from their employment and kept in isolation at home. The *Vektor* personnel were meanwhile subject to twice-daily checks for symptoms.[94] Two weeks after Vizunov was admitted to the isolation ward, he was "*practically dying*" and had a very high temperature. Vizunov's doctors undertook plasmapheresis on their patient and achieved a short-term improvement in his condition. The procedure was then repeated and a month later there were discussions regarding the possibility of a discharge. A second wave of illness then occurred and it was not until Vizunov had spent six months in hospital that he was eventually released on 14 August 1990.[95]

The case of Nikolai Vasil'evich Ustinov has close parallels with that of Vladimir Artemovich Pasechnik. In both cases, these were individuals engaged in advanced research with a burning desire to make a positive impact on the healthcare sector and thereby to improve the lives and well-being of their fellow citizens. Somehow, though, under the malign influence of the Soviet system, these individual aspirations became warped, misshaped and finally transfigured, bending the efforts of these talented scientists into much more malign objectives, the creation of a second generation of deadly offensive biological weapons. In the case of Ustinov, only his death enabled him to escape the overbearing pressure for perceived military advancement imposed by *Biopreparat*. Pasechnik, as will be documented below, was much more fortunate and his escape to the West facilitated his personal transformation into a fully-fledged biomedical researcher.

The Civil Cover Is Blown: The Defection of Vladimir Artemovich Pasechnik to the UK and the Exposure of *Biopreparat*'s Secret BW Research Programme

Biopreparat's R&D programmes were from the very outset subject to extraordinary levels of secrecy and compartmentalization. The vast bulk of both its headquarters staff, and the senior management in the BW research facilities themselves, were recruited from military biological institutions and related organizations such as the Soviet chemical troops. Their close familiarity with regimes of secrecy and their oaths of loyalty to the USSR and the military meant that remarkably scant human intelligence regarding *Biopreparat*'s secret BW effort ever reached the outside world. It was in fact one of the best-kept secrets of the Cold War. However, the monastic vows of silence so diligently adhered to by its staff were soon to be broken in the most dramatic of circumstances. For in October 1989 on a business trip to France, Dr Vladimir Artemovich Pasechnik, a high-level *Biopreparat* scientist, rang the British embassy in Paris, and became the most important source of information regarding the Soviet BW programme that the world had ever seen.

Such is the importance of Pasechnik—in Russian his surname means "beekeeper"—to this whole account, that it is necessary to trace his journey from humble beginnings to leading research scientist, to director of one of *Biopreparat*'s most important research facilities, the All-Union Scientific-Research Institute of Highly Pure Biopreparations—*Vsesoyuznyi nauchno-issledovatel'skii institut osobo chistykh biopreparatov*—abbreviated to *VNIIOChB* (Leningrad), and then to the role of international whistle-blower. Pasechnik was born on 12 October 1937 in Stalingrad, a city forever associated with the utmost horrors of the Second World War. Presumably the young Pasechnik was one of those fortunate enough to have been evacuated from the inferno in Stalingrad. During the intense fighting which engulfed the city from August 1942 to February 1943, more than 40,000 civilian inhabitants of the city are estimated to have been killed. However, the precise course of events at this crucial point in his life remains unknown.

We do know that Pasechnik's father, Artem Spiridonovich Pasechnik (see Fig. 6.4), born on 31 December 1913 into a peasant family in Ukraine, was in fact a genuine Soviet war hero. The son of a soldier who had died fighting for the White Army, he had originally been employed as an accountant in a village council. In 1937, the year of Pasechnik's birth, he had graduated from the Stalingrad Military Aviation Pilot School. He joined the CPSU in 1938 and was subsequently appointed as a squadron commissar. It was in this role that he then served in the Soviet-Finnish Winter War of 1939–1940. He flew several combat missions over Finland, destroying cars, locomotives and so on. On 11 March 1940, he was captured and later died following an aborted attempt to rescue his squadron commander and another colleague—one of the two men was a close personal friend—trapped on the ground. Artem Pasechnik was posthumously awarded both the title of Hero of the Soviet Union and the Order of Lenin. In 1982, a street in Leningrad was named in his honour.[96] Given the fact that his father had given his life in defence of the Soviet Union,

Fig. 6.4 Artem Spiridonovich Pasechnik, 14 November 1939 (Black Oak, Artyom Spiridonovich Pasechnik, Photo from the family album, 14 November 1939, https://commons.wikimedia.org/ wiki/File:Artyom_Spiridonovich_Pasechnik.jpg. This file is licensed under the Creative Commons Attribution-Share Alike 4.0 International license, Accessed on the 28 September 2020)

and the high esteem in which he was subsequently held by the state, it does make Vladimir Pasechnik's decision to defect, appear even more extraordinary.

Pasechnik was awarded a gold medal when he finished school. Later, he enrolled in the Lensovet Leningrad Technological Institute, where again his brilliance shone through. As noted previously, this higher educational institution appears to have administered a special programme to train specialists in chemical weapons production, but there is no evidence that Pasechnik was enrolled on any course of this type. It is reported that he studied biology and physics and rapidly became established as one of the institution's leading students.[97] After graduation, Pasechnik was employed as a researcher at the USSR Academy of Sciences' Institute of High Molecular Compounds in Leningrad. Here he worked under the internationally renowned scientist Professor Georgii Vasil'evich Samsonov who at this time was head of the institute's Department of Physiologically Active Polymers.[98] Pasechnik, who eventually headed his own laboratory, is reported to have been engaged in work on polymers with biological utility.

In 1974, Ogarkov, in his position as deputy director of *Glavmikrobioprom*, approached Samsonov with a request that he recommend one of his scientists to work on a new programme being pursued within the microbiological industry. With Pasechnik having been selected, Ogarkov went on to invite him to establish *VNIIOChB* in Leningrad, offering him a virtually unlimited budget with which to purchase lead-edge Western equipment and to recruit some of the best scientific staff in the Soviet Union. The new facility would form part of the All-Union Science Production Association *Biopreparat* headed by Ogarkov. The meeting between Ogarkov and Pasechnik appears to have taken place at the CPSU's City Committee in Leningrad. Apparently, amid appeals to his patriotic duty, his father's heroic record as a pilot was also alluded to.[99] Pasechnik appears to have been informed at this time of the intention to incorporate "closed" laboratories working on military projects at *VNIIOChB*. However, there was nothing particularly unusual in this situation in the USSR where so-called mobilization facilities were hidden within civil enterprises and many civil institutes incorporated military projects. Pasechnik, with huge ambitions to make an impact on Soviet healthcare, realized that this was a once in a lifetime opportunity, and quickly accepted Ogarkov's offer.[100]

VNIIOChB was officially created on 21 June 1974 under the terms of Order No. 189 issued by *Glavmikrobioprom*.[101] As was the case with *VNIIbiokhimmashproekt* (see above), the establishment of *VNIIOChB* was a complicated and lengthy process which was the focus of its new director from 1974 to 1981. In a similar fashion to *VNIIbiokhimmashproekt* (see above), the institute initially lacked any premises of its own and was temporarily housed in an old vacant building. At this time, *VNIIOChB* consisted of around 100 personnel working in just five laboratories. Construction of the first of the institute's three buildings began on 13 June 1978. Its main building with five floors and an area of 14,000 square metres was located in Leningrad on ulitsa Pudozhskaya, 7.[102] It is located a mere 1.25 kilometres north-east of the USSR Ministry of Defence's S.M. Kirov Military-Medical Academy. The new facility—commonly referred to as "*Pasechnik's institute*", is reported to have been completed at a cost of 5–7 million roubles and begun its work in 1981.[103]

By this time, Pasechnik, who is described as looking "*unlike a typical scientist ...short and stocky, with a round craggy face, a muscular build, and unusually beefy, artisan's hands*", had succeeded in recruiting around 400 personnel with an additional 200 support staff.[104] Remarkably little is known about the scientists based at this crucially important *Biopreparat* institute. Evgenii Nikolaevich Sventitskii was the deputy scientific director. He had graduated from the Physical Faculty of Leningrad State University in 1959. He was appointed head of a *VNIIOChB* laboratory in 1976. After Pasechnik's defection, he was initially appointed acting director and then confirmed permanently in this position in 1990.[105] Alibek reports another of the deputy directors in 1989 to have been one Nikolai Frolov but nothing else is known about him at all. Leitenberg and Zilinskas argue that unlike other facilities within the *Biopreparat* network (e.g. see the case of *VNIIbiokhimmashproekt* above), only two staff members were recruited from the military. One of these was a Colonel (Reserves) Evgennii I. Babkin who had apparently worked on *Y. pestis* at the USSR Ministry of Defence's Scientific-Research Institute of Microbiology (Kirov).[106] Babkin's position within *VNIIOChB* is not known but together with his other former military colleague, apparently revealed to their co-workers that they were the "*eyes and ears*" of Ogarkov.[107] It may be the case that given the extremely narrow focus of the institute, principally the development of formulations for weaponized agents, that staff from the military R&D institutes just simply did not possess the specialized competencies required for this task.

VNIIOChB is reported to have operated on an annual budget of 5 million roubles. This was supplemented with a hard currency allocation sufficient to purchase the latest scientific equipment from the West.[108] Some

85 per cent of the funding was directed towards R&D on biological weapons. One of the main tasks assigned to *VNIIOChB* was to develop the formulations and fillers for bacterial and viral pathogens weaponized at *Biopreparat*'s research centres in Obolensk and Kol'tsovo. Such work is difficult and can be of long duration. A key restraint facing the institute was that its location within a major city meant that the authorities would not grant permission for any work to proceed on Group 1 pathogens. The scientists worked around this obstacle by employing simulants in their work, thus using the non-pathogenic EV vaccine strain of *Y. pestis* for weaponized *Y. pestis* and so on.[109] The institute is reported at this time to have been in possession of its own large explosive chamber to test for dispersion and aerosol test rooms. Aerosol chambers are reported to have been installed both in the main *VNIIOChB* building and in its satellite laboratory housed in a dirty three-storey red-brick building at Korpusnaya ulitsa, 28. The latter was located a short distance from the main building on ulitsa Pudozhskaya, 7.[110]

From 1983 to 1985, *VNIIOChB* initiated its first scientific projects focused on the Group II pathogen, *Francisella tularensis*, the causative agent of tularaemia. The institute was then tasked with assisting in weaponization and subsequent formulation of the final weaponized product. However, as was the case with a similar project in Obolensk, the Leningrad scientists at first made no progress with regard to the weaponization of this agent. Later, a new research group based at Obolensk is reported to have succeeded in weaponizing *F. tularensis*. *VNIIOChB* is then reported to have developed the first principles for preparing a dry formulation of the weaponized agent which could be produced at the Omutninsk Chemical Factory.[111]

VNIIOChB also worked on the EV strain of *Y. pestis*. This work had the aim of developing a multi-antibiotic resistant strain and the creation of a dry formulation of *Y. pestis*. However, Pasechnik himself noted that there had been problems in the development of antibiotic-resistance, with each exposure to a different antidote making the bacteria become more feeble. He indicates that the problem was not resolved during his period as director, stating, "*The loss of biological activity during all technological processes was quite high, extremely high, I would say...this problem is quite complicated, in fact, and as far as I would know it wasn't solved efficiently*". [112] A similar lack of success was encountered with regard to creating a dry formulation of *Y. pestis*.

It is not known whether Pasechnik was at all involved in a secret project to create a new facility at Lakhta situated on the coast some 13.5 kilometres to the north-west of Leningrad. According to a 1992 US intelligence report, *VNIIOChB* was constructing, equipping and staffing a facility at Lakhta designed to do scale-up work to allow industrial production of a strain of plague—a strain developed to be resistant to cold and heat and to 16 antibiotics—for offensive purposes. The Lakhta facility was reportedly provided with a civil cover story—that it was engaged in research on vaccines and interleukin-2—to camouflage the military work.[113] This information appears to have been originally provided by a junior-level bench researcher based at *VNIIOChB* who defected to the UK in the Spring or early Summer of 1992 and was assigned the codename *Temple Fortune*. According to this source, *Biopreparat* was seeking to build a new facility at Lakhta for the production of a non-pathogenic EV vaccine strain of *Y. pestis* which could be substituted by a "*deadly, antibiotic-resistant form as soon as it was needed for weaponization*".[114]

On 24 August 1992, Douglas Hurd, the British Foreign Secretary and Lawrence Eagleburger, the US Secretary of State, sent a letter to the Russian Foreign Minister, Andrei Kozyrev. In it, they raised concerns that *Biopreparat*'s offensive programme was continuing and raised claims that "*a new plant was being built at Lakhta...for the large-scale production of germ warfare agents*".[115] Mark Urban, a leading UK defence correspondent, visited *VNIIOChB* with a BBC team in 1992. Here he was presented with photographs which showed that the Lakhta facility was in fact an unfinished shell.[116] Presumably the project was suspended sometime soon after Pasechnik's defection or in the wake of the collapse of the USSR.

A subsidiary task assigned to *VNIIOChB* concerned the synthesis of peptides and their testing in animal models to gauge their physiological effects.[117] This work formed part of a wider *Biopreparat* programme named *Factor* (see above) and focused on the potential military applications of various peptides, especially neuropeptides. The key unit working in this area at the institute was the Laboratory of Peptide Chemistry headed by Aleksandr Nikolaevich Kolobov. He had previously studied biochemistry at the A.A. Zhdanov Leningrad State University and graduated in 1982. He was appointed as a junior researcher at *VNIIOChB* in 1983. In 1993, Kolobov later emerged as a dynamic entrepreneur and he was one of the founders of the company, Verta Limited, which was a spin-off from *VNIIOChB* and was reported to employ "*two dozen former biological weapons scientists*". With help from the Civilian Research & Development

Foundation (CRDF) grants, Verta's researchers were at this time collaborating with the US biotechnology company SciClone to develop oral treatments for tuberculosis, hepatitis C and other diseases that are accompanied by suppression of the immune system.[118] This example vividly demonstrates, the potentially huge impact on the Soviet medical and health sectors which could have accrued from the redirection of *Biopreparat*'s weapons scientists to civil projects.

In 1985 in a remarkably ambitious *Biopreparat* project, Pasechnik's institute was ordered to begin work on the design of a modified cruise missile which would disperse BW aerosols via a spray system. The *VNIIOChB* scientists were specifically charged with modelling the process by which BW agents could be disseminated from a moving cruise missile. Such a weapon, if it penetrated enemy air defences, could cause potential havoc in a targeted country. Pasechnik went on to defect while the project was still in its early stages. Leitenberg and Zilinskas draw on other testimony to suggest that the project had not reached the stage of completion when the Soviet Union collapsed.[119]

The research capabilities of *VNIIOChB* were highly appraised by the *Biopreparat* management. Alibek, who was first deputy director of *Biopreparat*, regarded the institute as one of his most important facilities and notes that it "*provided many of our breakthroughs in weapons production*".[120] He reports that one of the facility's most notable projects concerned the creation of a milling machine which used "*a powerful blast of air to turn bacterial and viral mixtures into a fine powder*".[121] It was intended that this technology would replace existing ball-bearing-based mills at military facilities and be introduced at *Biopreparat* plants. Alibek also refers to work completed by *VNIIOChB* on drying and microencapsulation employing polymer capsules.

Mark Urban, who was the first journalist to interview Pasechnik in the UK in 1993, provides a fascinating glimpse into the mindset of the *VNIIOChB* staff. He notes that they developed a considerable degree of cynicism, particularly among those "*who had previous experience of expensive but unproductive military research in the Soviet system. At the same time, scientists at the institute were well treated by Soviet standards. The result was mixed feelings typical of the Soviet 'era of stagnation'. People used to say, 'We're involved in a highly-paid job, there's no reason to change it…but at the same time there were sad feelings about the uselessness of this programme and its vicious characteristics*'".[122]

It is interesting to note that on 15 January 1985, Anthony Rimmington, the author of this current work, and at that time a postgraduate student at the University of Birmingham's Centre for Russian and East European Studies (CREES), made an application to the British Council for a research visit to the USSR. The application was made under the terms of the existing Anglo-Soviet Cultural and Scientific Agreement and proposed a programme of work focusing on Soviet microbiology and the industrial production of microbial proteins in the USSR. Rimmington requested that a number of meetings be held with scientists working on Soviet biotechnology programmes including those based at *VNIIOChB* in Leningrad.[123] The British Council reported in October 1985 that Rimmington's name was not on the list of accepted scholars which it had received from the Soviet authorities.[124] It appears likely that there was alarm among Soviet security agencies that the UK postgraduate could somehow gain knowledge of *Biopreparat*'s offensive BW projects being pursued at this time at the institute.

The appointment on 11 March 1985 of the reforming Mikhail Gorbachev as General Secretary of the CPSU encouraged Pasechnik to seek to change the direction of research being pursued by his institute. Beginning in late 1985, he began attempts to pressure senior *Biopreparat* officials to allow him to allocate resources towards civil objectives. This effort appears to have eventually met with some partial success at least. In September 1987, Pasechnik was appointed General Director of the newly created Science Production Association *Farmpribor* (*NPO Farmpribor*) which incorporated his own facility, two additional institutes and two manufacturing plants with a total of 2500 employees.[125] *VNIIOChB* was the lead institute of the new association. Another component part of *Farmpribor* was the *OKB TM* design institute in Kirishi which focused on the design of equipment for the manufacture and dispersal of biological weapons.[126] Another identified unit within the association was the Leningrad-based Scientific-Research Institute of Vaccines and Sera (*NIIVS*). This facility had originally been founded in 1948 and during the prolonged siege of Leningrad in the Second World War, had remained operational, delivering its preparations both within the city and to Red Army units. From 1970 to 1976, a Special Science and Production Complex was constructed for *NIIVS* at a 15-hectare site in the Krasnoe Selo district of Leningrad. Upon completion the institute was relocated to the new site. There is no known linkage to any Soviet BW activity at the institute which appears to have focused on the creation of preparations for the civil sector. *NIIVS* at this time incorporated its own Enterprise for the Production of Bacterial Preparations which focused on the manufacture of vaccine preparations.[127] Leitenberg and Zilinskas report that the newly created *Farmpribor* was totally oriented towards the civil sector. This was useful from the authorities' point of view in that it further obscured the military programmes being actively pursued by *VNIIOChB*. In addition, sales of its vaccines and other civil products generated vital income streams for *Biopreparat* at a time when the USSR was in the midst of a growing political and economic crisis. Despite its privileged status, this wider crisis eventually impacted *VNIIOChB* and in 1987 its budget for BW was reduced. It was then forced to generate some 30 per cent of its income from civil consulting activities and the sale of its own products.

By now, Pasechnik occupied a number of key positions within *Biopreparat* as a whole, sitting on its Scientific and Technical Council as well as its Council of Directors and heading the association's relations with the USSR Academy of Sciences. *Biopreparat* also had its own civil space biology programme and from 1988 he was also its representative on a national space committee. He had established close personal relations with key scientists working within the network including, for example, Vladimir Petrovich Zav'yalov, who was director of the Institute of Immunology (Lyubuchany). Moreover, he had made "*multiple official visits to various Biopreparat facilities for over a dozen years*".[128] Domaradskii, who was based at *VNII PM* (see above) in Obolensk, reports that Pasechnik frequently accompanied a commission which "*almost every month*" visited his facility. The commission was headed by Kalinin and also included representatives of the *VPK* and the CPSU Central Committee.[129] All this provided Pasechnik with an overarching view of the association's entire network and the ambitious offensive BW programme which it was pursuing. In late 1984, he attended a meeting of *Biopreparat*'s Scientific Council, which endorsed the programme of work for *Biopreparat* from1985 to 1990. This new programme embraced work on the development of antibiotic-resistant strains of *Y. pestis* (described above).[130]

Pasechnik appears to have been driven by the desire to do everything he could to be considered worthy in the eyes of his father, the Soviet war hero. As a consequence, he was firmly focused on the personal contribution he could himself make, using the application of lead-edge molecular biology, to create new drugs and treatments for the health service. Therefore, the realization that *VNIIOChB* had been selected to play such a pivotal and substantive role in the Soviet Union's offensive BW programme had come as a great shock and caused him to grow increasingly anxious. By the late 1980s, he is reported to have been going through a severe personal crisis, suffering from constant depression, nervous breakdowns and sleepless nights.[131] Pasechnik himself reports that he experienced "*feelings like misunderstanding, anxiety and then disgust—disgust towards this programme and to himself and to the people who have been involved in this programme*".[132] Writing in *Izvestiya* on 26 June 1993, the Russian journalist S. Leskov provides a penetrating analysis of Pasechnik's state of mind at this moment. Commenting on his subsequent defection, he states: "*...such a sharp turn of fortunes of Pasechnik is interesting in itself. As the son of a hero of the Soviet Union, he had become the director at the age of thirty-eight and he had excellent prospects for further promotion. But...having realised the real objects of the programme in his own words he had decided to reveal them to the world*".[133]

Alibekov appears to have caught a glimpse of Pasechnik's state of mind at this time and noted during a meeting with him in Leningrad in 1988 that he appeared tired and a little depressed. After the meeting Pasechnik drove Alibek to a railway station to catch the sleeper-train back to Moscow. It was during the drive that Pasechnik sadly commented, "*Its like this, I'm fifty-one years old, and I'm going through a strange time in my life. I don't know if I've accomplished what I want to. And they're going to make me retire soon*".[134] Pasechnik was aware that he faced mandatory retirement from his position at the age of 55 and the bleak prospect of surviving on a barely adequate pension of 120 roubles a month.

It was against this complex background, embracing both his own personal health and imminent diminution in professional status, that, by 1988, Pasechnik had become firmly focused on attempting an escape from the clutches of the system operated by *Biopreparat*. Overseas travel presented the best opportunity for Pasechnik to implement his plans. He had already had some limited experience of such trips, having as director of *VNIIOChB*, visited the Soviet Union's Warsaw Pact ally, Czechoslovakia. He had been accompanied by Lieutenant-General Vladimir Lebedinskii, from the USSR Ministry of Defence's Fifteenth Administration. The two scientists inspected Czech aerosol test chambers which they were interested in acquiring for their own facilities.[135]

There is wide divergence in the accounts of what happened next. Leitenberg and Zilinskas, who interviewed the Russian scientist most recently, report that in 1988 Pasechnik travelled with Kalinin, head of *Biopreparat*, to a biotechnology exhibition in West Germany. Pasechnik apparently already had engaged in some preliminary negotiations concerning a joint project with a German company which were to be followed up. In addition, Kalinin viewed this event as an opportunity for *Biopreparat* to pursue the creation of new commercial links with other Western companies.[136] The international trip may have been highly convivial for the two scientists since they were, according to Domaradskii, very close work colleagues and good friends.[137] Despite his high rank, Alibekov appears to have been totally unaware that Pasechnik had undertaken this trip to the West with Kalinin.

The opportunity to escape that Pasechnik had been waiting for then presented itself towards the end of October 1989. During a visit to Leningrad, Alibekov was approached by the *VNIIOChB* director who was seeking permission to visit an industrial exhibition in Toulouse, France. Here the latter was scheduled to attend, together with a head of one of his laboratories, a meeting with a French equipment manufacturer who had a new line of fermenters. Alibekov acceded to his request. Domaradskii expresses his amazement that Pasechnik was ever able to undertake this visit, stating, "*To this day I am mystified as to how Pasechnik found himself abroad; it must have been an oversight of our Secret Office*".[138] Pasechnik managed to give the KGB minder and other members of his party the slip and then took a train to Paris. Here he is reported to have approached the Canadian embassy but officials there refused to believe his account.[139] He had more success with the British embassy and after several hours of unavoidable delay, which involved deliberation by UK intelligence agencies and the procurement of the requisite authorizations from various ministries, he was quickly picked up and flown to London.

Pasechnik—who was now assigned the codename *TRUNCATE*—was met upon arrival in London by a small group of senior officials from the British Secret Intelligence Service (SIS), MI6. He was then transferred to a safe house outside the capital. In November an intensive series of debriefings began which was conducted by the UK Defence Intelligence Service (DIS). Brian Jones, who had a doctorate in metallurgy, led the DIS unit, DI-53, concerned with analysis of nuclear, biological and chemical weapons intelligence data. He was aided by only two CBW experts including Surgeon Commander Christopher Davis, known as 53b. Jones therefore requested that additional technical assistance during the debriefing be provided by the Chemical Defence Establishment, and David Kelly, CDE's senior principal scientific officer and others were seconded to the team that debriefed Pasechnik over the next three years or so. Early in 1990, the information gleaned during the process was passed to the CIA and, later, intelligence officers from the US were invited to meet with Pasechnik.[140] Davis believes, "*The fact that Vladimir defected was one of the key acts of the entire ending of the Soviet Union and the end of the Cold War. It was the greatest breakthrough we ever had*".[141]

The Soviet government is reported to have made a sustained effort to have Pasechnik returned to the USSR. The Soviet ambassador in London who had complained that Pasechnik was being held against his will was allowed by MI6 to talk with Pasechnik over the telephone. The ambassador apparently promised Pasechnik that "*there would be no problems and all his concerns would be fully addressed*". However, Pasechnik's response was to write a long letter to the Soviet government outlining his reasons for defecting and resigning from the Communist Party.[142]

Meanwhile, back in the USSR, a team from *Biopreparat*'s headquarters is reported to have been despatched to Leningrad where they secured all the documentation in Pasechnik's office. A thorough search was implemented in order to ascertain precisely what information he had had access to and what he might reveal to UK intelligence. All of the classified materials stored at *VNIIOChB* were systematically destroyed at this time with rooms full of documents being shredded and burned.[143] The KGB placed officers in Pasechnik's apartment to be ready in case he should attempt to contact his family.[144] A damage-control team was also created which held meetings in Kalinin's Moscow office. A decision was quickly taken to destroy any documentation relating to secret work undertaken by *VNIIOChB* and to redirect work at the institute to fully civil objectives. As has been discussed, there had already been some impetus in this direction but it was recognized that suspension of military work in Leningrad would have left a major hole in *Biopreparat*'s offensive programme.[145]

After his debriefing, Pasechnik was provided with a detached house purchased for him in Shrewton, a village located on Salisbury Plain. He was then employed for the next nine years or so by the Centre for Applied Microbiology and Research (CAMR), at Porton Down, which was at that time part of the Department of Health's Public Health Laboratory Service. Here he continued an active career as a research scientist. Meanwhile, after quiet negotiations with the Soviet authorities, Natasha, his wife, joined him in the UK. Later, his son Nikita and his own family also settled in Shrewton.[146] However, Pasechnik, quite understandably, never stopped looking over his shoulder and was concerned that the KGB and their successors would attempt to kill him. This was one of the reasons he cut himself off from old friends.[147]

Pasechnik appears to have rapidly achieved a satisfactory level of personal happiness after his defection to the UK. Christopher Davis reports that during his last debriefing session with Pasechnik "*he was greeted with a huge warm smile and an enthusiastic hand-shake. 'The twinkle had returned to his eye', he recalls. They went to a pub together, and shared a meal and a pint or two. Later they went for a walk through the meadows surrounding the town…Pasechnik was relaxed, happy and chatted easily*".[148] His son, Nikita, reports that Pasechnik also became close friends with a fellow biological weapons expert, David Kelly, who had been part of the original briefing team.[149] However, the Russian scientist was much less happy in his new role at CAMR, where as an outsider, he was confined to the margins and really only used as an all-purpose problem solver. Clearly, given his background deep within the Soviet offensive BW complex, there would inevitably be questions continually raised by his colleagues and others about how far Pasechnik could be trusted.[150]

It was at this juncture in his life that Pasechnik had the good fortune to meet with a life sciences entrepreneur from Canada named Caisey Harlingten. The latter had spent many years studying alternative therapies to antibiotics for bacterial infections and had focused on phage therapies. Pasechnik is reported to have become excited about the potential of this approach and proposed to Harlingten that they join forces to create a new company. It was after this, that in 1999, Pasechnik left CAMR and on 2 January 2000 formed a new company called Regma BioTechnologies. Pasechnik was appointed Chief Scientific Officer of the new company which initially was housed in a temporary lab rented from CAMR. Later, a facility designed to his precise specifications was built. During the period through to November 2001, he filed numerous patent applications which formed the basis of the company's scientific and commercial development. Regma's principal therapeutic target was tuberculosis. This is described as a happy time for Pasechnik who at long last was pursuing his life-long passion for creating cures for use against human diseases.[151] A year or so after creating Regma, Pasechnik had to undergo a quadruple bypass operation. Within a month he appeared to have fully recovered and was working on designing the necessary steps to take his phage treatment to full clinical testing. Pasechnik subsequently suffered a major stroke and collapsed at his home. He was to die three weeks later in hospital on 21 November 2001.[152] He was buried in Shrewton and was survived by his wife, Natasha, two sons and a stepdaughter.[153]

Notes

1. Istoriya v dokumentakh, https://www.obolensk.org/about.htm, Accessed on 17 September 2020.
2. *Professora voenno-meditsinskoi (mediko-khirugicheskoi) akademii (1798–1998)*, Nauka, St Petersburg, 1998, p. 98.
3. Information provided to Dr Anthony Rimmington (University of Birmingham) during a visit to Obolensk, 7 June 1993.
4. Information provided to the author by former *Biopreparat* scientist, UK, February 2007; Domaradskij, I., Orent, W., *Biowarrior*, p. 198.
5. The Soviet archives indicate that in 1986, the general contractor for construction of production areas at *VNII PM* was the USSR Ministry of Medium Machine Building (*Minsredmash*) and the subcontactor was the USSR Ministry of Defence. See Report of the Main Administration *Biopreparat* on Capital Investment for Industrial Enterprises and Organizations for 1986 and Explanatory Notes, *RGAE*, Fond 788, Opis' 1, Delo 99, p. 22.
6. Information provided to the author by former *Biopreparat* scientist, February 2007.
7. Report of the Main Administration *Biopreparat* on Capital Investment for Industrial Enterprises and Organizations for 1986 and Explanatory Notes, p. 22.
8. Nedogonov, V., Plague in our house, *Soviet Weekly*, No. 2591, 31 October 1991, p. 10.
9. Observations made during a visit to Obolensk by Dr Anthony Rimmington (University of Birmingham), 7 June 1993.
10. Domaradskij, I., Orent, W., *Biowarrior*, p. 186.
11. Ibid., p. 199.
12. Rimmington, *Who's Who*, p. 259.
13. Domaradskij, I., Orent, W., *Biowarrior*, p. 200.
14. Ibid.
15. Ibid., p. 207.
16. Ibid., p. 209.
17. Ibid., p. 213.
18. Nikolai Nikolaevich Urakovu 60 let, *Biotekhnologiya*, Vol. 5, No. 6, p. 785.
19. Rimmington, *Who's Who*, pp. 299–300.
20. Domaradskij, I., Orent, W., *Biowarrior*, pp. 213–222.
21. Ibid., pp. 223–227.
22. Ibid., pp. 216–217.
23. Urakov, N.N., Obolensk Scientific-Production Complex: Outlook for Cooperation with Western Companies, p. 2, Transcript of presentation

made at *Russian Biotechnology—An Overview of its Standards and Efficiency and Opportunities of Cooperation with Western Companies*, A workshop organized by the Pharmaceutical Worldwide Database, DECHEMA Building, Frankfurt, Germany, 1–2 March 1993. The author attended this event and presented a copy of his latest book to Urakov. The latter reciprocated by presenting a typewritten copy of the transcript of his presentation.

24. Leitenberg, Zilinskas, with Kuhn, *The Soviet Biological Weapons Programme*, p. 588.
25. Domaradskij, I., Orent, W., *Biowarrior*, p. 235.
26. Central Collection of Microorganisms of State Research Center for Applied Microbiology, The OECD Megascience Forum, Unique Research Facilities in Russia, Vol. 2, http://web.archive.org/web20001212161900/http://www.oecd.org/dsti/sti/s_t/ms/prod/russia/1ch6.htm, Accessed on 28 September 2020.
27. Laboratory Complex of P-3 Level of Physical Protection for Studies on the Agents of Severe Infections Static-Climatic; Horizontal Dynamic Device (SC-10; HDD-600), The OECD Megascience Forum, Unique Research Facilities in Russia, Vol. 2, http://web.archive.org/web/20001212153100/http://www.oecd.org/dsti/sti/s_t/ms/prod/russia/1ch5.htm, Accessed on 28 September 2020.
28. Domaradskij, I., Orent, W., *Biowarrior*, p. 247.
29. Ibid., p. 246.
30. Leitenberg, Zilinskas, with Kuhn, *The Soviet Biological Weapons Programme*, pp. 702–703.
31. Miller, J., Engelberg, S., Broad, W., *Germs: Biological Weapons and America's Secret War*, Simon & Schuster, New York, 2001, p. 208.
32. Interview—Dr Serguei Popov, 1 November 2000, *Journal of Homeland Security*, 13 November 2000, www.homelandsecurity.org, Accessed 23 September 2003.
33. Ashmarin, I.P., Regulyatornye peptidy dlya meditsiny, *Fundamental'nye nauki—meditsine, Sovmesnaya sessiya Obshchego sobraniya AN SSSR i Obshchego sobraniya AMN SSSR 19–20 noyabrya 1980*, Nauka, Moscow, 1981, pp. 201–203.
34. Ibid.
35. Interviews with Biowarriors: Sergei Popov, Nova Online, September 2003, http://www.pbs.org/wgbh/nova/bioterror/biow_popov.html, Accessed on 30 May 2005.
36. Leitenberg, Zilinskas, with Kuhn, *The Soviet Biological Weapons Programme*, pp. 194–195.
37. Hoffman, D.E., *The Dead Hand*, p. 299.
38. Miller, J., Engelberg, S., Broad, W., *Germs*, pp. 300–304.

39. Prizes, http://vector.sibinfo.ru/DesktopDefault.aspx?lcid=25&tabid=21&tabindex=1, Accessed on 6 May 2004.
40. Rimmington, *Who's Who*, pp. 179–180.
41. Sagdeev, R.Z., *The Making of a Soviet Scientist.*
42. Ob organizatsii problemnogo nauchno-proizvodstvennogo otdela po razrabotke i proizvodstvu novykh biologicheskikh preparatov-gipnoproteinov, Ministerstvo meditsinskoi promyshlennosti SSSR Prikaz No. 139-a, 13 April 1990, *RGAE*, Fond 788, Opis 1, p. 1341.
43. O sozdanii malogo gosudarstvennogo predpriyatiya "Neiropeptid" (MGP "Neiropeptid"), Ministerstvo meditsinskoi promyshlennosti SSSR, Prokaz No. 388, 21 November 1990, *RGAE*, Fond 788, Opis 1, p. 1347.
44. O vkhozhdenii MGP "Neiropeptid" v sostav Vsesoyuznyi assotsiatsii "Farmbiopress", Ministerstvo meditsinskoi promyshlennosti SSSR, Prikaz No. 132, 10 April 1991, *RGAE*, 788, Opis' 1, p. 1407.
45. Bykov, V.A., O pereimenovanii Lotoshinskogo biokhimicheskogo zavoda, Ministerstvo meditsinskoi promyshlennosti SSSR Prikaz, No. 212, 8 June 1990, *RGAE*, Fond 788, Opis 1, Delo 1342, p. 136.
46. Ratnikova, L.I., Vypiska iz istorii Lotoshinskogo biokhimicheskogo zavoda, RSFSR Lotoshino Regional Council of Peoples Deputies, Executive Committee, Riagosarkhiv, No. 46, 28 March 1990, *RGAE*, Fond 788, Opis 1, Delo 1342, p. 137.
47. Chuprin, V., We Are Undoubtedly the First. Anthrax Vaccine Is Made Only in... Lotoshino, *Moskovskii komsomol'ets*, 16 February 2002, pp. 1–2.
48. Istoriya Tsentra, http://www.vector.nsc.ru/o-gnts-vb-vektor/, Accessed on 24 September 2020.
49. Chelovek, davshii imya poselku, *Naukograd-Vesti*, No. 12(177), 13 July 2012, p. 3, https://kolcovo.ru/content/news/vesti/177_min.pdf, Accessed on 24 September 2020.
50. Odarenko, V., "Vektor" i Kol'tsovo; vekhi stroitel'stva, *Naukograd-Vesti*, No. 10(223), 30 May 2014, p. 6, https://kolcovo.ru/content/news/vesti/223_min.pdf, Accessed on 25 September 2020.
51. Istoriya Tsentra, http://www.vector.nsc.ru/o-gnts-vb-vektor/, Accessed on 24 September 2020.
52. The unusual nature of the activity at both Kol'tsovo and *Akademgorodok* appears to have been identified very rapidly by Western intelligence agencies with information being released into the open media as early as 1984. This noted the existence of a new institute "*prominently located near the village of Bareshevo along the much-travelled road between Novosibirsk and a huge scientific research complex called Academe City*". See Kucewitz, W.,

The Science of Snake Venom, *The Wall Street Journal*, 25 April 1984, p. 30.
53. Chelovek, davshii imya poselku, p. 3.
54. Odarenko, V., "Vektor" i Kol'tsovo; vekhi stroitel'stva, p. 6.
55. Pervye ob"ekty i sobytiya v Kol'tsovo: kak vse nachinalos', https://www.naukogradpress.ru/glavnoe/2020/09/19/pervye-obekty-i-sobytiya-v-koltsovo-kak-vse-nachinalos/, Accessed on 24 September 2020.
56. "Vektor": 45 let na strazhe biobezopasnosti, https://www.naukogradpress.ru/glavnoe/2019/08/29/vektor-45-let-na-strazhe-biobezopasnosti/, Accessed on 24 September 2020.
57. Marakhovskaya, I., Pervye ob"ekty i sobytiya v Kol'tsovo; kak vse nachinalos', https://www.naukogradpress.ru/glavnoe/2020/09/19/pervye-obekty-i-sobytiya-v-koltsovo-kak-vse-nachinalos/, Accessed on 12 October 2020.
58. Rimmington, *Who's Who*, pp. 152–253.
59. Pamyati L.S. Sandakhchieva, http://www.vector.nsc.ru/pamyati-lssandahchieva/, Accessed on 25 September 2020.
60. "Vektor": 45 let na strazhe biobezopasnosti, https://www.naukogradpress.ru/glavnoe/2019/08/29/vektor-45-let-na-strazhe-biobezopasnosti/, Accessed on 24 September 2020.
61. Vspominaya L'va Sandakhchieva, *Naukograd-Vesti*, No. 12(177), 13 July 2012, p. 4, https://kolcovo.ru/content/news/vesti/177_min.pdf, Accessed on 24 September 2020.
62. Leitenberg, Zilinskas, with Kuhn, *The Soviet Biological Weapons Programme*, pp. 209, 784 (note 9).
63. Vorob'ev, A.A., *Dvadtsatyi vek moimi glazami*, Medical Information Agency, Moscow, 2005, p. 191.
64. Sandakhchiev, L.S., *Innovation Development Programme of the Association "VECTOR" for 1992–1995 and till (sic) 2005*, State-Owned Concern "BIOPREPARAT" SCIENTIFIC-PRODUCTION ASSOCIATION "VECTOR", Novosibirsk, 1991, p. 3.
65. Alibek, K, with Handelman, S., *Biohazard*, pp. 117–118.
66. Sandakhchiev, L.S., *Innovation Development Programme of the Association "VECTOR" for 1992–1995 and till (sic) 2005*, p. 3.
67. Ibid., p. 18.
68. Ibid., pp. 4–7.
69. Odarenko, V., "Vektor" i Kol'tsovo; vekhi stroitel'stva, p. 6.
70. P-4 Biocontainment Virological Laboratory Experimental Building, The OECD Megascience Forum, Unique Research Facilities in Russia, Vol. 2, http://web.archive.org/web/20001212085700/http://www.oecd.org/dsti/sti/s_t/ms/prod/russia/1ch1.htm, Accessed on 25 September 2020.

71. Unique Climatic Static-Dynamic Unit UKSD-25, The OECD Megascience Forum, Unique Research Facilities in Russia, Vol. 2, http://web.archive.org/web/20001212144000/http://www.oecd.org/dsti/sti/s_t/ms/prod/russia/1ch3.htm, Accessed on 25 September 2020.
72. Ibid.
73. Leitenberg, Zilinskas, with Kuhn, *The Soviet Biological Weapons Programme*, p. 213.
74. Laboratory Animals Farm, The OECD Megascience Forum, Unique Research Facilities in Russia, Vol. 2, http://web.archive.org/web/20001212133900/http://www.oecd.org/dsti/sti/s_t/ms/prod/russia/1ch2.htm, Accessed on 25 September 2020.
75. Alibekov, K.I., Prilozhenie No. 1 k prikazu Ministerstva meditsinskoi promyshlennosti SSSR, No. 122, 5 April 1990, *RGAE*, Fond 788, Opis 1, Delo 1341, pp. 9–10.
76. Ibid., p. 5.
77. *Belorusskii ordena trudovogo krasnogo znameni nauchno-issledovatel'skii institut epidemiologii i mikrobiologii (BelNIIEM)*, Minsk, 1997, p. 5.
78. Information provided by scientists within Group of Molecular Biology and Biotechnology, Scientific-Research Institute of Epidemiology & Microbiology, Minsk, 18 February 1997.
79. Van der Groen, G., *Bibliography: 1966–2019*, Institute of Tropical Medicine, Antwerp, 18 January 2019, http://lib.itg.be/guidovandergroen_bibliography.pdf, Accessed on 28 September 2020.
80. Kuhn, J.H., *Filoviruses: A Compendium of 40 Years of Epidemiological, Clinical, and Laboratory Studies*, Springer-Verlag, Wien, 2008, p. 54, https://doi.org/10.1007/978-3-211-69495-4, Accessed on 28 September 2020.
81. The Belarusian account is corroborated by Sergei Netesov, Director of the Institute of Molecular Biology, Kol'tsovo, Novosibirsk *oblast'*. He reports that the State Scientific Centre of Virology & Biotechnology *Vektor* received strains of Ebola and Marburg from the Scientific-Research Institute of Epidemiology & Microbiology. *ProMED report*, Vol. 95, No. 178, 29 November 1995, reproduced on the "Outbreak" home page at http://www.outbreak.org.
82. Leitenberg, Zilinskas, with Kuhn, *The Soviet Biological Weapons Programme*, p. 216.
83. Rimmington, *Who's Who*, pp. 249–250.
84. Leitenberg, Zilinskas, with Kuhn, *The Soviet Biological Weapons Programme*, pp. 218–219.
85. Ibid., p. 192.
86. Rimmington, *Who's Who* (1994), p. 252.

87. Dolzhenkova, V., Tomskii sled "brata" Eboly, *Tomskie novosti*, 7 November 2014, http://tomsk-novosti.ru/tomskij-sled-brata-eboly/, Accessed on 29 September 2020.
88. Ibid.
89. Raskryty smerty uchenykh-virusologov v tsentre "Vektor", Angarskii Portal Novostei, 19 February 2020, https://xn%2D%2D38-6kcaak9aj5chl4a3g.xn%2D%2Dplai/raskryty-smerti-uchenyh-virusologov-v-centre-vektor/.
90. Ibid.
91. Alibek, K, with Handelman, S., *Biohazard*, p. 131.
92. Raskryty smerty uchenykh-virusologov v tsentre "Vektor", Angarskii Portal Novostei, 19 February 2020, https://xn%2D%2D38-6kcaak9aj5chl4a3g.xn%2D%2Dplai/raskryty-smerti-uchenyh-virusologov-v-centre-vektor/.
93. Turovskii, D., "Drugogo sposoba lecheniya net i seichas", *Meduza*, 31 October 2014, https://meduza.io/news/2014/10/31/drugogo-sposoba-lecheniya-net-i-seychas, Accessed on 29 September 2020.
94. Ibid.
95. Raskryty smerty uchenykh-virusologov v tsentre "Vektor", Angarskii Portal Novostei, 19 February 2020, https://xn%2D%2D38-6kcaak9aj5chl4a3g.xn%2D%2Dplai/raskryty-smerti-uchenyh-virusologov-v-centre-vektor.
96. Geroi strany, http://www.warheroes.ru/hero/hero.asp?Hero_id=2421, Accessed on 20 August 2020.
97. Vladimir Pasechnik, *The Telegraph*, https://www.telegraph.co.uk/news/obituaries/1363752/Vladimir-Pasechnik.html, Accessed on 21 April 2005.
98. Georgii Vasil'evich Samsonov (5.09.1920–25.06.99), Sorbtsionnye i khromatograficheskie protsessy, Vol. 11, No. 6, 2011, p. 740, http://www.sorpchrom.vsu.ru/articles/20110601.pdf, Accessed on 19 August 2020.
99. Nikita Pasechnik. Solsberi. Zhizn' i smert' otsa, http://drugoivzgliad.com/nikita_pasechnik_salisbury/, Accessed on 20 August 2020.
100. Adams, J., *The New Spies: Exploring the Frontiers of Espionage*, Hutchinson, London, 1994, p. 270.
101. Official website of the Federal Medical-Biological Agency's Federal State Unitary Enterprise "State Scientific-Research Institute of Highly Pure Biopreparations", https://hpb.spb.ru/ru/institute/history/, Accessed on 20 August 2020.
102. Ibid.
103. Adams, J., The Red Death: The Untold Story of Russia's Secret Biological Weapons, *The Sunday Times*, 27 March 1994, News Review, Section 4,

p. 1; Leitenberg, Zilinskas, with Kuhn, *The Soviet Biological Weapons Programme*, p. 251.

104. Mangold, T., Goldberg, J., *Plague Wars*, p. 91.
105. Sventitskii represented *VNIIOChB* as acting director at a meeting of a Soviet-Swiss pharmaceutical working group on 5 June 1990. See Telegin, L.P., O pooshchrenii rabotnikov VNIIOChB, PO "Oktyabr'", VNITIAF, USSR Ministry of the Medical Industry, Order No. 202, 5 June 1990, *RGAE*, Fond 788, Opis 1, Delo 1342, p. 106. Confirmation of his position as director is found in Gosudarstvennyi nauchno-issledovatel'skii institut osobo chistykh biopreparatov, https://rucompany.ru/boss.php?id_company=1863&id_boss=1893, Accessed on 21 August 2020.
106. Leitenberg, Zilinskas, with Kuhn, *The Soviet Biological Weapons Programme*, p. 252.
107. Ibid., pp. 252, 792.
108. Ibid., p. 271.
109. Ibid., p. 255.
110. Cooper, S., Life in the Pursuit of Death, *Seed*, No. 4, January/February 2003, p. 71.
111. Leitenberg, Zilinskas, with Kuhn, *The Soviet Biological Weapons Programme*, p. 255.
112. Urban, M., *UK Eyes Alpha: The Inside Story of British Intelligence*, Faber and Faber, London, 1996, p. 131.
113. An unsigned document, found in State Department records and according to them dating from September 1992, summarizes information the anonymous author had been gathering for years from former Soviet citizens who had been involved with the Soviet biological warfare programme since the 1970s. Memorandum re Soviet Biological Weapons Programs, no classification, 17 September 1992 (according to source), 10 pp., https://nsarchive2.gwu.edu/NSAEBB/NSAEBB61/Sverd32.pdf, Accessed on 24 August 2020.
114. Mangold, T., Goldberg, J., *Plague Wars*, pp. 118–140. For more on Temple Fortune, see also Hoffman, D.E., *The Dead Hand*, pp. 429–430.
115. Urban, M., *UK Eyes Alpha*, pp. 226–227.
116. Ibid., pp. 227–228.
117. Domaradskij, I., Orent, W., *Biowarrior*, p. 206; Leitenberg, Zilinskas, with Kuhn, *The Soviet Biological Weapons Programme*, p. 255.
118. Star Wars Into Ploughshares, *Economist.com*, 15 March 2005, http://www.economist.com/science/displayStory.cfm?story_id=3715108, Accessed on 15 March 2005.
119. Leitenberg, Zilinskas, with Kuhn, *The Soviet Biological Weapons Programme*, p. 255.
120. Alibek, with Handelman, *Biohazard*, p. 140.

121. Ibid.
122. Urban, M., *UK Eyes Alpha*, pp. 131–132.
123. Anglo-Soviet Cultural and Scientific Agreement Form of Proposal, Anthony Rimmington, 15 January 1985.
124. Scholarship to the Soviet Union, Letter (SOV/172/2A) from Amanda Alabaster (Overseas Educational Appointments Department, The British Council) to Anthony Rimmington, 2 October 1985.
125. *NPO Farmpribor* was created on 17 September 1987 under the terms of Order No. 730 of the USSR Ministry of the Medical and Microbiological Industry. Istoricheskaya spravka, *RGAE*, Fond 788, Opis' 1.
126. Lücke, E-M., Poetzsch, E., *Biotechnology Directory Eastern Europe*, Walter de Gruyter, Berlin, 1983, p. 226.
127. Rimmington, A., *The Biotechnology and Pharmaceutical Industry in Russia 2010*, p. 86.
128. Leitenberg, Zilinskas, with Kuhn, *The Soviet Biological Weapons Programme*, p. 584.
129. Domaradskij, I., Orent, W., *Biowarrior*, p. 209.
130. Urban, M., *UK Eyes Alpha*, p. 131.
131. Pasechnik, N., Solsberi. Zhizn' i smert' otsa, http://drugoivzgliad.com/nikita_pasechnik_salisbury/, Accessed on 20 August 2020.
132. Urban, M., *UK Eyes Alpha*, p. 129.
133. Domaradskij, I., Orent, W., *Biowarrior*, p. 272.
134. Alibek with Handelman, *Biohazard*, pp. 139–140.
135. Leitenberg, Zilinskas, with Kuhn, *The Soviet Biological Weapons Programme*, p. 466.
136. Ibid., pp. 256–257.
137. Domaradskij, I., Orent, W., *Biowarrior*, p. 272.
138. Ibid., p. 281.
139. Pasechnik, N., Solsberi. Zhizn' i smert' otsa.
140. Mangold, T., Goldberg, J., *Plague Wars*, p. 91.
141. Some corroboration of Pasechnik's account was subsequently provided by another Soviet defector who, in the Autumn of 1990, sought asylum at the British Embassy in Helsinki. See Hoffman, D.E., *The Dead Hand*, pp. 322 and 353.
142. Mangold, T., Goldberg, J., *Plague Wars*, pp. 100–101.
143. Ibid., p. 100.
144. Bannerman, L., Sergei Skripal: Salisbury's other spy lived in fear of KGB revenge, *The Times*, 12 March 2018, https://www.thetimes.co.uk/article/city-s-other-spy-lived-in-fear-of-kgb-revenge-fhkwhnvbp, Accessed on 26 August 2020.
145. Alibek with Handelman, *Biohazard*, p. 144.
146. Bannerman, L., *The Times*, 12 March 2018.

147. Ibid.
148. Cooper, S., Life in the Pursuit of Death, p. 106.
149. Nikita Pasechnik, Solsberi, Zhizn' i smert' otsa.
150. Cooper, S., Life in the Pursuit of Death, p. 106.
151. Ibid., pp. 106–107.
152. Ibid, p. 107.
153. Vladimir Pasechnik: Defector who alerted the west to the danger of Soviet biological weapons production, *The Guardian*, 28 November 2001, https://www.theguardian.com/news/2001/nov/28/guardianobituaries.highereducation, Accessed on 26 August 2020.

CHAPTER 7

A Roadmap to the Future? The Emergence of *Biopreparat* as a Major Civil Biopharmaceutical Player

The election in March 1985 by the Politburo of Mikhail Gorbachev as General Secretary of the CPSU was to have a profound impact on the politics and economics of the Soviet Union. The first stage of his *perestroika* (reconstruction) project was focused on *uskorenie* (acceleration) and *glasnost'* (openness). Gorbachev initially believed that these policies could be combined with enhanced economic growth and ambitiously aimed to match the industrial output of the US by 2000. However, it soon became apparent that radical reform of the economic system and the dislocation which resulted were likely to reduce, rather than increase, production in the short term.[1] It should also be stressed that Gorbachev's reforms did not aim at any dismantlement of the centrally planned economy.

It was against this background of rapid economic and political change that the Central Committee of the CPSU issued Decree No. 1118 on 14 November 1985, and the Presidium of the USSR Supreme Soviet issued Order No. 3612-XI on 21 November 1985, regarding the merging of *Glavmikrobioprom* (the Main Administration of the Microbiological Industry under the USSR Council of Ministers) with the USSR Ministry of the Medical Industry.[2] The two bodies were now to form the newly created USSR Ministry of the Medical and Microbiological Industry (*Ministerstvo meditsinskoi i mikrobiologicheskoi promyshlennosti SSSR* or *Minmedbioprom*).[3] *Glavmikrobioprom* contributed highly significant resources to the new combined ministry, incorporating by this time, well

A. Rimmington, *The Soviet Union's Invisible Weapons of Mass Destruction*, https://doi.org/10.1007/978-3-030-82882-0_7

over 100 factories, 4500 staff working within research facilities and another 3500 specialists employed within design institutes.[4] One of the driving forces for the reorganization was the recognition by the Soviet leadership that the medical industry was wedded mainly to traditional techniques and lacked any knowledge or familiarity with the cutting-edge technologies in molecular biology which were beginning to have such a huge impact in the West.

Glavmikrobioprom had already demonstrated its capabilities in the healthcare area to the Soviet leadership. During the CPSU's 25th Congress in 1976, Anatolii Petrovich Aleksandrov, the President of the USSR Academy of Sciences, had reported on a successful project to develop a chromatographic inactivated influenza vaccine. The work had been conducted by three facilities in Leningrad, the Pasteur Institute of Epidemiology and Microbiology, the M.I. Kalinin Polytechnical Institute and the B.P. Konstantinov Institute of Nuclear Physics. Aleksandrov immediately came under pressure from the party leadership to organize the industrial production of the new preparation.[5] Kosygin is then reported to have issued a decree for production of the influenza vaccine to be established at *Biopreparat*'s Omutninsk Chemical Factory (Vostochnyi, Kirov *oblast'*, see above). The decree set an output target of 20 million doses per annum within a three-year time frame. This project, which was initiated in 1979, required the conversion of part of the Omutninsk factory and installation of new equipment, since up to this point the plant had been focused only on bacterial products and had no experience of producing viral preparations. Belyaev, the head of *Glavmikrobioprom*, charged Vorob'ev (First Deputy Scientific Director of *Biopreparat*) and Valov (Director, Omutninsk Chemical Factory) to head this work. The factory was rebuilt over a period of six to eight months with the help of virologists from the USSR Academy of Medical Sciences. They acted quickly to establish large-scale facilities for embryonated chicken eggs in which the virus could be propagated. By 1980, some 8 million doses of vaccine had been produced by the Omutninsk factory.[6] In his memoirs, Vorob'ev claims that his team met the specified target of 20 million doses capacity by the end of year three. However, the Soviet archives reveal that even as late as 1986, production at the site had only reached 17.9 million doses.[7] Tests conducted by the L.A. Tarasevich State Scientific-Research Institute for the Standardization and Control of Medical Biological Preparations confirmed the Omutninsk vaccine to be of a very high quality. Valov was awarded the title of Hero of Socialist Labour as a result.[8] Vorob'ev was awarded the USSR State Prize for his role in 1980.[9] Interestingly, the Soviet archives reveal that a few years later in 1985, a mechanized line for sorting and rejecting chicken eggs and embryos was installed at the Omutninsk factory—presumably this technology also formed part of the influenza vaccine project.[10]

Biopreparat's Omutninsk Chemical Factory also played an important role in another civil endeavour, the Soviet *Revertaza* project. This was launched in 1975 and aimed to isolate and purify reverse transcriptase (revertase). This enzyme was considered to be essential for work undertaken in molecular biology and in particular for the cloning of genes. The *Revertaza* project concentrated on the isolation and purification of revertase from the avian myeloblastosis virus (AMV). It is reported to have incorporated the work of specialists from many different fields including molecular biology, enzymology, virology and genetics. Scientists from Novosibirsk, Riga, Kiev, Moscow, the German Democratic Republic and Czechoslovakia participated in the project.[11] The technology developed by the scientists working on the *Revertaza* project was then transferred to the Omutninsk Chemical Factory, which in 1985 launched production of an RNA-dependent DNA polymerase preparation from avian myeloblastosis virus. *Biopreparat*'s All-Union Scientific-Research Institute of Applied Microbiology (*VNIIPM*) at Obolensk (Moscow *oblast'*) is reported to have played a key role in setting up production of the enzyme at the factory.[12]

One of the main stages in the enzyme production process is the extraction of purified viral concentrate from the plasma of chicks with acute leukaemia. Up until 1989 at least, ultrafiltration was used for this purpose. However, its drawbacks included the high cost of ultracentrifuges, their limited throughput, and the attendant difficulties of scale-up. In 1986, workers at *VNIIPM* explored the possibility of using Soviet-produced hollow fibres for concentration and purification of avian myeloblastosis virus under industrial conditions. They found that it was possible to use these fibres for scaling up isolation of the myeloblastosis virus with no necessity to increase the number of centrifuges.[13] Soviet scientists are reported to have been particularly proud of the success of the *Revertaza* project (1975–1979) and several State Prizes were awarded to those involved. However, it should be pointed out that series production of reverse transcriptase at the Omutninsk Chemical Factory did not begin until the beginning of 1985, some five years after the official "*success*" of the programme.[14]

It was not only at Omutninsk where *Biopreparat* facilities were making a significant impact on the civil sector. In 1980, *VNIIOChB* (Leningrad, see above) working under the leadership of Vladimir Pasechnik had begun work on the study of cytokines, their role in regulatory processes and the creation of drugs based on them. The institute subsequently developed the Soviet Union's first technology for the production of highly purified leucocytic human interferon *Interlok*. This technology was then transferred to *Biopreparat*'s Vyshnii Volochek Factory of Enzyme Preparations (Kalinin *oblast'*) where the preparation was produced on an industrial scale. This work of the *VNIIOChB* team which led the project was recognized with the award of Soviet orders and medals.[15]

The Omutninsk influenza vaccine and reverse transcriptase projects, and *VNIIOChB*'s success with *Interlok*, had clearly raised the profile of *Biopreparat* among the senior Soviet leadership and provided a vivid demonstration of its potential capabilities in the civil arena. Therefore, when *Minmedbioprom*, the new ministry, was created, the All-Union Science Production Association was granted the higher status of a main administration and henceforth was known as the Main Administration *Biopreparat* (*Glavnoe upravlenie "Biopreparat"*). It coexisted with 19 other administrations including the First Administration (*Pervoe upravlenie*), staffed by the KGB, which controlled security. Interestingly, *Biopreparat* was charged at this time with acting on a *khozraschet* (self-financing) basis, which was an integral part of the Gorbachev policy of perestroika.[16]

There appears to be little doubt that *Glavmikrobioprom* and *Biopreparat* were given the lead role in the new super-ministry. The minister of *Minmedbioprom* was none other than Valerii Alekseevich Bykov, who had been appointed head of *Glavmikrobioprom* in 1985. Prior to that he had served from 1971 to 1976 as director of the Kirishi *Biokhimiya* plant which had a projected capacity of 100,000 tonnes per annum of microbial protein using petroleum derivatives as feedstock. His background within the CPSU had assisted him in his career progression. He had joined the party in 1966, then served as the First Secretary of the Kirishi City Party Committee and had also been a delegate to the 25th CPSU Congress in February 1976. In 1979, he secured his most important party appointment as head of the Microbiological Industry Section of the CPSU Central Committee's Chemical Industry Department.[17]

Another strong pointer to the priorities of the new ministry can be gleaned by the sections established in late March 1986 under its

Scientific-Technical Council. These included: No. 1—Molecular biology, genetics and selection of industrial microorganisms; No. 2—Biotechnology and technological equipment for medical and microbiological production; No. 3—Biological engineering, mechanization and automation for production processes; No. 4—Immunology and allergology; No. 5—Production hygiene, technical safety, preservation of natural resources and environmental protection; No. 6—Industrial economics, scientific-technical prognoses and so on; No. 7—Synthetic drugs and vitamins and their nomenclature; and No. 8—Antibiotics and other drugs based on natural raw materials and their nomenclature.[18] During the period 1986–1987, the council was comprised of some 86 persons, of whom 26 were representatives of the USSR Academy of Sciences, the State Committee for Science and Technology, the State Planning Committee, the USSR Ministry of Health, the State Agro-industrial Committee, the Ministry of Higher Education and other interested ministries and departments. A total of 6 academicians, 23 Doctors of Science and 17 Candidates of Science sat on the council. It should perhaps come as no surprise that the deputy chairman of *Minmedbioprom*'s Scientific-Technical Council at this time was none other than Major General (Reserves) Dmitrii Vladimirovich Vinogradov-Volzhinskii, a key representative of the USSR Ministry of Defence in the *Biopreparat* system.[19]

The head of the new Main Administration *Biopreparat* was Lieutenant General (Reserves) Yurii Tikhonovich Kalinin who also served as a deputy minister of *Minmedbioprom*. He had originally been appointed to head *Biopreparat* in 1980 and at that time was appointed concurrently as first deputy head of *Glavmikrobioprom*. Several different accounts point to family links assisting in his rise to prominence. Domaradskii claims that rumours circulating at the time ascribed his appointment to the fact that his wife's family had "*close ties with powerful people in high places*".[20] Alibek also points to his wife as exerting influence as the daughter of a four-star general. Mangold and Goldberg, meanwhile, report, "*His rise to power began when he left his first wife, the daughter of a laboratory director, to marry the daughter of a senior general in the Soviet Defence Ministry*".[21] Whatever the case, there can be no doubting his abilities and Domaradskii acknowledges that "*Kalinin is a very bright man with a good feel for political conditions, and he is well able to adjust and adapt quickly to new situations*".[22]

Kalinin is credited with being an "*empire builder*" and increasing the size, scope and political importance of *Biopreparat*.[23] Much of this expansion appears to have taken place after the USSR Council of Ministers issued a decree (10 December 1987) which was followed on 15 January 1988 by a corresponding *Minmedbioprom* Order No. 18 which created a much slimmer central apparatus for the Ministry of the Medical and Microbiological Industry. *Biopreparat* now constituted one of only 12 administrations within *Minmedbioprom*.[24] However, it was the subsequent orders issued by the Ministry which are of far greater significance. For, on 5 February 1988, five major new science production associations (so-called *NPOs*—see above, which embraced R&D institutes and factories) were created under *Biopreparat*. These included *NPO Allergen* (based in Stavropol'), *NPO* Nutrient Media (based in Makhachkala, Dagestan), *NPO* Bacteriophage (Tbilisi, Georgia), *NPO Biomed* (Perm') and *NPO Immunopreparat* (Ufa, Bashkortostan).[25] By now, *Biopreparat* controlled around 40 facilities in 15 different cities, giving Kalinin control of the country's most important vaccine R&D institutes and production plants together with a crucial manufacturer of nutrient medium. Given that virtually the entirety of its original focus had been on military projects, *Biopreparat* had now surprisingly also emerged as the most important player in the Soviet Union's civil biopharmaceutical sector.

It is worth examining the composition of just one of the science production associations transferred to *Biopreparat*, *NPO* Bacteriophage. It comprised of the Tbilisi Scientific-Research Institute of Vaccines & Sera together with a major new vaccine production plant which was built in 1974. The plant, the unimaginatively named Enterprise for the Production of Bacterial Preparations, was situated on territory immediately adjacent to its parent research institute. By 1983 production of dry *STI* anthrax vaccine in the Tbilisi plant stood at 3,420,000 ampoules (with each ampoule containing 20 doses of vaccine).[26] Interestingly, following the accidental release in April 1979 of an anthrax aerosol from the USSR Ministry of Defence's BW facility in Sverdlovsk, there was an intensive vaccination campaign within the city. Around 80 per cent of the 59,000 inhabitants of the Chkalovskii District (close to the military facility) who were considered eligible were vaccinated with at least one shot of the *STI* vaccine produced at the Tbilisi Institute of Vaccines & Sera.[27] However, no *STI* vaccine was sent from Tbilisi to Ekaterinburg in the aftermath of the incident and the local population were presumably vaccinated with preparations stockpiled by the USSR Ministry of Health or the military.[28] It is apparent from this one example that *Biopreparat* was at this time absorbing facilities of critical importance to the Soviet health system.

Another area of Soviet civil bioscience where *Biopreparat* was emerging as a major player concerned the breeding and supply of laboratory animals for research. On 28 July 1988, Bykov issued *Minmedbioprom* Order No 491 "*On the creation of a branch system of breeding centres for laboratory animals*". Under this directive, Urakov (see above) was to head a new organization which it was envisaged would eventually manage and coordinate the breeding and supply of lab animals at sites across the USSR. Bykov's order also called for profitable sub-departments for breeding laboratory animals to be established both at *NPO Biosintez* (Obolensk) and at *NPO Vektor* (Kol'tsovo). At the latter, Sandakhchiev (see above) was charged with calculating the projected cost of creating a large-scale nursery for breeding large animals such as horses, miniature horses, pigs and mini pigs. Sandakhchiev was to be provided with the necessary funding and assisted by *VNIIbiokhimmashproekt* with the design and equipping of such a facility, with construction scheduled to begin in 1990. Kalinin was given overall responsibility for the implementation of this order.[29]

Meanwhile, in Moscow, the relentless reorganizations of the Soviet biopharmaceutical and medical sectors continued apace. On 27 June 1989, under the terms of a USSR statute, *Minmedbioprom* was reformed as the Ministry of the Medical Industry (*Ministerstvo meditsinskoi promyshlennosti SSSR* or *Minmedprom*) with Bykov retaining his position as minister. Kalinin, as head of *Biopreparat*, served concomitantly as a deputy minister of *Minmedprom* with an official salary of 800 roubles per month.[30] There were also changes implemented with regard to the central apparatus of the new ministry. On 29 September 1989, *Minmedprom* issued Order No. 384a which transformed the Main Administration *Biopreparat* into the State Concern *Biopreparat* (*Gosudarstvennyi kontsern "Biopreparat"*) with the organization remaining under the control of the new ministry.[31] The bureaucratic changes are important since they appear to represent a major tilt in orientation by *Biopreparat* towards the civil sector. There appears to have been no significant diminution in its offensive BW programme as is evidenced by the testimony of Pasechnik—the work for the military carried on unabated. Rather, *Biopreparat*'s absorption of a large number of civil biopharmaceutical institutes and production plants meant, according to Alibek, that by 1990 the "*supervision of vaccine production and antibiotic development for our state health system occupied almost 50 per cent of our official functions*".[32] Kalinin and some senior staff members are reported to have had no interest in these civil facilities, but Alibekov, the first deputy director, and presumably others within the organization, found themselves being increasingly drawn to close involvement in this aspect of *Biopreparat*'s activities.

Engagement with the West: The Emergence of Kanatzhan Baizakovich Alibekov

In the late 1980s, Kanatzhan Baizakovich Alibekov (see Fig. 7.1) was emerging as a highly influential figure within *Biopreparat* and the wider Soviet life sciences sector. At this point in time, the relatively young scientist occupied the position of first deputy chairman of the board of the State Concern *Biopreparat (Pervyi zamestitel' predsedatelya Pravleniya kontserna "Biopreparat")*.[33] In that position he represented *Biopreparat* on a number of important committees, including an expert group focused on drafting plans for the development of medical and microbiological industries in the next decade and a half.[34] In February 1990, Alibekov was rewarded with a bonus of 550 roubles for his role in assisting *NPO Vektor* and the Enterprise *Antigen* organize the development and industrial production of HIV diagnostics.[35] On 11 April 1990, he was given joint responsibility for implementing a ministerial order regarding the use of new influenza strains in the manufacture of vaccines and diagnostics.[36] A month later, in May 1990, he

Fig. 7.1 Kenneth Alibek (the former Kanatzhan Baizakovich Alibekov), First Deputy Chairman of the State Concern *Biopreparat* (USG: NIH, Dr Ken Alibek, former Soviet biowarfare expert, http://www.nih.gov/catalyst/2003/03.05.01/Alibek.jpg, 1 May 2003, Accessed on the 18 August 2020. This work is in the public domain in the United States because it is a work prepared by an officer or employee of the United States Government as part of that person's official duties under the terms of Title 17, Chapter 1, Section 105 of the US Code)

was also appointed chair of the organizing committee for an important meeting on the supply and use of laboratory animals, both within *Biopreparat* and more widely in the USSR Ministry of the Medical Industry.[37]

At some point early in 1990, Kalinin, in conjunction with the Soviet authorities, selected Alibekov to be the lead figure in engagement with the West with regard to both commercial and defence-related biological issues. Given his in-depth knowledge of the offensive biological warfare activities underway within the ostensibly civil organization, this does appear to be an extraordinary decision. It may be the case that the defection a few months earlier to the UK of Vladimir Pasechnik, a high-level *Biopreparat* scientist, meant that the authorities were more relaxed about allowing Western access to Alibekov. After all, the West by now would already have had a fairly detailed knowledge of the network. It is also the case that Alibekov was atypical of senior *Biopreparat* personnel in that he had never served within the highly secret BW facilities attached to the USSR Ministry of Defence and had no detailed knowledge of the programmes in which they were engaged. He may therefore have been considered "low-risk" as far as the Soviet intelligence services were concerned.

Alibekov's emergence as a key figure in *Biopreparat* responsible for engaging with Western companies may also be connected with the proven entrepreneurial flair he had exhibited during negotiations with a local vodka distillery. The *Kristall* (Crystal) distillery was in fact the immediate neighbour of *Biopreparat*'s headquarters building on Samokatnaya ulitsa. At some point around 1989, Alibekov had used their canteen when his own had been closed for renovation. Here he had a chance encounter with one of the managers at *Kristall*. The latter reported that he was seeking to acquire land suitable for the construction of a plant for the manufacture of plastic miniature vodka bottles which could be served on flights. Brands to have been bottled in this way were to have included "Stolichnaya" and "Golden Ring". Alibekov suggested that the Soviet concern utilize territory belonging to *Biopreparat*'s Institute of Immunology at Lyubuchany for construction of a new facility.

One of the first indications of Alibekov's new internationally focused role, gleamed from official documents held within the Russian State Archive of the Economy (*RGAE*), was when, in August 1990, he provided a spoken report to the USSR Ministry of the Medical Industry's Scientific-Technical Bureau Council on his discussions with Eli Lilly (US) and Novo Nordisk (Denmark) for the construction of turn-key factories in the Soviet Union for the production of genetically engineered human insulin.[38] Alibekov's report later refers to the following companies as being interested in cooperation with the Soviet Union in this area: Eli Lilli and Upjohn (both US), Novo Nordisk (Denmark), Hoechst (West Germany) and Biobraz (Brazil).[39]

At this period in time, Alibekov was also to play a pivotal role in *Biopreparat*'s close interaction with a small UK-based company, Oxford Virology plc (OV). The latter was founded in 1986 and possessed a workforce comprising of only a dozen employees. The company was best known for the work of its consultants, Professor David Bishop (NERC Institute of Virology and Experimental Biology, Oxford) and Chil-Yong Kang (University of Ottawa), who developed recombinant strains of baculovirus both as insecticides and as gene expression systems. It was also engaged in work on vaccine candidates and diagnostics for HIV, haemorrhagic fever and bluetongue, as well as diagnostics for hepatitis B. Early in 1990, OV was quoted on the now-defunct Third Market for junior companies and succeeded in raising over £2 million. However, by 1991 it had become apparent that OV remained three to five years away from realizing the commercialization of its proprietary biological control products, and it was actively seeking to identify new opportunities for generating more immediate revenue streams.[40]

The interaction between OV and *Biopreparat* was very much science-driven and is reported to have arisen after a delegation of eight leading Soviet molecular biologists visited the UK in July 1989.[41] Subsequently, Professor Bishop attended a scientific meeting in Moscow and appears to have met with a senior scientist from the Science Production Association *Vektor—NPO Vektor* (Kol'tsovo, Novosibirsk *oblast'*). It rapidly became apparent that "*there was scope for collaboration with the Soviet organization*".[42] In 1990, in what appears to be the first such visit of its kind, members of OV and the Institute of Virology met in Kol'tsovo arranged by *Vektor*'s Director Professor Lev Stepanovich Sandakhchiev.

A key British participant in the meeting was Derek Lennon, Chairman of OV. He was able to point to the fact that, during the 1960s, he had sold and organized the construction of very large petrochemical complexes in the Soviet Union.[43] Mr Lennon noted during his visit that "*Vektor is really a notable scientific organisation, with its own independent achievements in genetic engineering and virology. I was also very impressed at their ability in modelling and computer methods in molecular biology and genetics*".[44] A number of commercial agreements then followed, under the terms of which the UK-based company secured 50 per cent of the rights to innovations developed at *Vektor*'s All-Union Scientific-Research Institute of

Molecular Biology and it also licensed three compounds developed by this institute which had been shown to have potential utility in the treatment of AIDS. One of the treatments, Ridostine, subsequently underwent initial trials at King's College Hospital in London. If successful, OV had intended approaching a Western pharmaceutical company to undertake clinical trials. A brochure detailing the full range of products and services offered by *Vektor* was also advertised and distributed by Oxford Virology in the early 1990s.[45]

John Hodgson, a leading journalist working at Nature's *Bio/Technology* journal was sceptical as to the successful outcome of the new venture. He pointed out: "*The deal is steeped in insecurity: uncertainties surrounding the quality of Siberian products coupled with OV's inexperience in marketing. NPO will supply products on consignment despite the absence of specified targets for OV, a reflection of both parties' need for hard currency. Oxford Virology's marketing manager, Richard Ackers, also hopes to soon sign up two other smaller Soviet manufacturing organizations under similar terms*".[46]

It was sometime around this point that Alibekov rose to prominence in the discussions with OV. It appears highly likely that Kalinin had been informed of the interaction and had subsequently instructed him to become involved as the representative of the *Biopreparat* management. In the face of the ongoing political and economic collapse of the Soviet Union, it appears that this was considered to be a major opportunity for *Biopreparat* to access lucrative international markets. Alibekov contacted Derek Lennon and proposed that OV form a joint venture with both *Biopreparat* and *NPO Vektor* which would be focused on marketing. On 5 December 1990, the following telex message was sent to OV by Alibekov:

> *Attention Mr Derek Lennon of Oxford Virology.*
>
> *Dear Mr Derek Lennon, Thank you very much for your work done. I keep receiving from you very important and useful information. Under current situation* (sic) *the most profitable action is considered to be setting up a joint venture (JV) comprising Oxford Virology, the Moscow Narodny Bank, Concern Biopreparat, and NPO Vector. The JV's job will have to be marketing services, promoting into the world market the products and technologies of Biopreparat subordinate facilities and research and development centres.*
>
> *I deem it necessary to start preparation of foundational documents and to discuss same during your next visit to Moscow. On behalf of Concern Biopreparat I am ready to grant a general written mandate for Oxford Virology to represent the Biopreparat interests and to conduct appropriate negotiations with the Moscow Narodny Bank.*

> *Please keep me informed regarding your progress in talks with the bank, preparation of basic foundational papers on establishing JV, English draft of the mandate as you see it, as well as your plans with respect to your coming visit dates and signing the documents.*
>
> *Yours truly Alibekov*
> *05 December 1990*[47]

On 14 December 1990, OV reported that Alibekov had approved the creation by the UK company of a database of *Biopreparat* products. Alibekov reported that *Biopreparat* at that time controlled 36 factories with an annual turnover of Roubles 800 million (£800 million at official exchange rates).[48] As part of the proposed deal, *Biopreparat* also offered OV exclusive rights with regard to an oil-degrading bacterium which it had developed at its facilities in Novosibirsk.[49] Unfortunately for Alibekov, OV ceased trading prior to the completion of negotiations after being defrauded by one of their UK distributors.[50] However, this case is another vivid illustration of the direction *Biopreparat* was heading at this time. It was clearly emerging from the shadows as both a national and an international player in the biopharmaceutical sector with its eyes fixed on potentially huge commercial rewards.

This was a hectic period for Alibekov. Just over three weeks after sending the telex to OV, on 30 December, he was appointed concomitantly as the director of *NPO Biomash* (see above).[51] However, despite his desire to direct his efforts towards civil projects, Kalinin now selected him as the ideal candidate to front *Biopreparat* in its interactions with a visiting UK-US inspection team. During 1990–1992, both the US and the UK governments had put considerable diplomatic and political pressure on Gorbachev to "*come clean*" on the Soviet offensive BW programme. In response, the Soviet leader invited a 12-member UK-US team to visit the four most important *Biopreparat* R&D facilities at Lyubuchany, Obolensk, Kol'tsovo and Leningrad. The inspection began on 8 January 1991 with a visit to the Institute of Immunology (later to be renamed as the Institute of Engineering Immunology or *III*—see below).[52] Although Alibekov had previous experience in interacting with Western companies, he lacked at this time, any knowledge of English.[53] The inspection team were denied access to the dynamic aerosol test chambers and plague research laboratories at Obolensk. A tussle with a Soviet official also occurred when Christopher Davis, a UK representative (see above) attempted to turn on a flashlight within a large freestanding hexagonal steel chamber at the site.

Eventually, the official relented and Davis was able to discover evidence that biological bombs had been exploded inside the chamber.[54] At Kol'tsovo, on 14 January, a member of the technical staff informed David Kelly, another UK representative (see above), that his laboratory was working on smallpox. The World Health Organization had previously eradicated the disease and samples of the virus were only authorized to be held at two official repositories, at the Centres for Disease Control (Atlanta, US) and at the Scientific-Research Institute of Viral Preparations (Moscow).[55] Senior officials at *Vektor* later contradicted this admission. They were, however, unable to account for "*the research being undertaken in a dynamic aerosol test chamber on orthopoxvirus, which was capable of explosive dispersal*".[56] The final visit by the inspection team was to Pasechnik's former institute in Leningrad (see above). Despite the fact that the former director of *VNIIOChB* had defected and been intensively debriefed in the UK, his former colleagues were still prepared to argue that explosive test chambers were being used in agricultural projects and that contained jet-stream milling machines were used "*for the grinding of salt*". According to David Kelly, "*Candid and credible accounts of many of the activities at these facilities were not provided*".[57]

Alibekov's emerging entrepreneurial skills were again highlighted in the summer of 1991 when, as director of *Biomash*, he interacted with a US businessman, Joel Taylor, who had an interest in exporting used hospital equipment to the USSR.[58] Later, in December 1991 Alibekov participated as the most senior ranking representative of *Biopreparat* in a 13-man Soviet delegation which made reciprocal visits to biological installations in the US. On 13 January 1992, shortly after the collapse of the USSR, Alibekov resigned from his positions at *Biopreparat*.[59] Later in 1992, having already established contacts in Washington, he travelled to the US and took the new name of "Ken Alibek". In the period through to the end of 1993, he was then debriefed by senior officials from US intelligence agencies and a range of government departments.[60]

Alibekov then is a crucial figure in the history of *Biopreparat*, since he was probably the first senior manager to fully recognize and embrace a new direction of travel for the organization. This was to include a major refocus on civil projects with the development of products for the Russian health service. There was also to be a heavy emphasis on commercial activity, including interaction with international biopharmaceutical companies. It is clear from the evidence presented above that just prior to the collapse of the USSR, Alibekov was heavily engaged in *Biopreparat* projects which had these objectives very much in mind. Much is also owed to Alibekov/Alibek for the incredible insight he has provided with regard to the history, development and management of *Biopreparat*. His unique knowledge, with regard to the creation of new BW pilot facilities at Berdsk, Omutninsk and Stepnogorsk, based on his own leadership and participation in these projects, is especially critical in our understanding of the evolution of the *Biopreparat* network.

Decay, Corruption and Environmental Protest: The Birth of the State Concern *Biopreparat*

In Alibek's description of *Biopreparat*'s operations in the last two years or so prior to the collapse of the USSR, one receives a very strong sense of the decay and corruption which was evidenced on a daily basis at this time in the Soviet Union. In February 1990, for example, it was discovered that *Biopreparat*'s vaccine plant in Tbilisi (Georgia)—see Fig. 7.2—was

Fig. 7.2 View of Fermenter at *Biopreparat*'s former Enterprise for the Production of Bacterial Preparations, Tbilisi Georgia, 1 November 2000. (©A. Rimmington)

supplying non-sterile vaccine vials to local hospitals. Alibek, together with Valerii Ganzenko, head of the Medical Administration, undertook an inspection visit to the facility and discovered that *Biopreparat* funds had been misappropriated by the local director and poorly paid staff were having to work with 40-year-old equipment.[61] This case was representative of a much wider malaise in the Soviet health sector. In 1989, for example, *Biopreparat* had shipped boxes of disposable plastic syringes to medical facilities across the country. This was in response to 250 children having been infected in the main paediatric hospital in the Soviet city of Elista after using contaminated syringes. Nurses pointed to the lack of equipment and staff which prevented them from using adequate sterilization methods.[62]

Another key feature of the wider microbiological industry at this time was the rise of a nationwide environmental protest movement against discharges from plants belonging to *Minmedbioprom/Minmedprom.* The first evidence of a major crisis came with reports in the Soviet press in 1988 of appalling environmental pollution caused by the large-scale microbial proteins factory in Kirishi, near Leningrad (see above). The plant, which used petroleum derivatives as feedstock, had been built at a cost of 180 million roubles in 1975.[63] However, local doctors were soon linking repeated releases of protein dust to a 20-fold increase in the number of cases of bronchial asthma in the town.[64] Several children were reported to have died as a result and more than 100 people made permanent invalids.[65] Martin Walker, writing in the British newspaper, *The Guardian*, described how the situation in Kirishi had deteriorated into "*the worst ecological disaster since Chernobyl*". [66] There were repeated attempts by the local authorities to shut the plant down but the minister, Bykov, who had previously been director of the plant, refused to budge and production was resumed.[67] In the wake of the Kirishi affair, the Soviet media, given new freedoms by Gorbachev's policy of *glasnost*, reported on numerous accidents and discharges of harmful pollutants at Soviet biotech plants across the country. As with the nuclear industry in the West, the biotech industry became synonymous with hazardous pollution in the Soviet consciousness. Local authorities acted to shut down several factories, and public rallies succeeded in preventing construction of new plants. By the late 1980s, it had become practically impossible to build a biotech facility anywhere in the Soviet Union.[68]

Biopreparat was not immune from the public backlash. The situation at the Production Association *Progress* in Stepnogorsk typified the public perception of the microbiological industry at this time. In April 1991, it was reported that as a result of continual discharges and a constant odour, the local population was becoming increasingly opposed to the *Biopreparat* facility. The association's management exacerbated the already difficult situation by refusing to allow representatives from the local committee on the environment to monitor the release of harmful substances at the plant.[69] *Biopreparat* also became the target of a vociferous local popular campaign directed against the potential environmental impact of a viral production plant, which in 1987 was being constructed at a remote site in Mochalishche in the Yoshkar-Ola region.[70] A special commission was subsequently appointed in response to the campaign and declared the project unfeasible. As a result, the 300 million roubles (around US$400 million) project was mothballed and then later abandoned.[71]

Just immediately prior to the collapse of the USSR, there was to be one final organizational reshuffle. The USSR Cabinet of Ministers issued Decree No. 170-50 on 4 April 1991 under the terms of which some 14 institutes, 23 factories, two pilot plants, one design bureau and assorted other facilities were transferred to the State Concern *Biopreparat* (see Appendix C).[72] As in the previous reorganization, these included all the core BW facilities and some of the country's most important vaccine facilities. In addition, however, *Biopreparat* was to take control of the Soviet Union's major producers of antibiotics. This included two plants, in Penza and Kurgan, which are reported to have incorporated reserve mobilization BW capacities (see above). Meanwhile, just a few days later, on 13 April 1991, the USSR Cabinet of Ministers issued Decree No. 176 which liquidated the USSR Ministry of the Medical Industry.[73] Possibly, the nationwide campaign against the microbiological industry influenced the Soviet authorities in this decision. *Biopreparat* had for many years been the tail that was wagging the dog. Now it had dramatically thrown off the shackles of being subordinate to a higher ministry and emerged as a self-managing independent state concern. Kalinin had clearly been active behind the scenes in ensuring his concern exited the ministry in possession of most of its prized biopharmaceutical assets. The reach and power of his organization had risen to its zenith just as the dissolution of the Soviet Union was to take place and the frantic struggle over control of *Biopreparat*'s enormously valuable assets was to begin.

Notes

1. Brown, A., Kaser, M., Smith, G.S., *The Cambridge Encyclopedia of Russia and the Former Soviet Union*, Cambridge University Press, Cambridge, 1994, p. 338.
2. V prezidiume verkhovnogo soveta SSSR, *Pravda*, 22 November 1985, p. 4.
3. Istoricheskaya spravka, *RGAE*, Fond 788, Opis' 1.
4. Rychkov, R.S., Mikrobiologicheskaya promyshlennost'—Meditsine, in *Fundamental'nye nauka-meditsine, (Sovmestnaya sessiya Obshchego sobraniya AN SSSR i Obshchego sobraniya AMN SSSR, 19–20 noyabrya 1980g.)*, Nauka, Moscow, 1981.
5. Neiman, B.Ya., *Industriya mikrobov*, Znanie, Moscow, 1983.
6. *Fundamental'nye nauki—meditsine, (Sovmestnaya sessiya obshchego sobraniya AN SSSR i obshchego sobraniya AMN SSSR 19–20 noyabrya 1980g.)*, Nauka, Moscow, 1981.
7. Report of the Main Administration "Biopreparat" on Basic Activities of Industrial Enterprises for 1986 and Explanatory Notes, *RGAE*, Fond 788, Opis' 1, Delo 98, p. 56.
8. Vorob'ev, A.A., *Dvadtsatyi vek moimi glazami*, pp. 210–212.
9. Vorob'ev, A.A., *Ne podvodya cherty: Nauchno-publististicheskaya povest'*, Medical Information Agency, Moscow, 2003, p. 411.
10. Reports on *Biopreparat*, *RGAE*, Fond 178, Opus 1, Delo 5436–5439.
11. Lutz, Y., Graf, H., Canonne, P., *Biotechnologie en URSS: Recherche et developpement*, La Recherche Scientifique et Technologique Dans Le Monde, No.14/SGDN/STS/VST/4, Premier Ministre, Secretariat General de la Defense Nationale, 3rd June, 1988.
12. Vnimaniyu organizatsii, *Biotekhnologiya*, No. 5, 1985; *Biotekhnologiya*, No. 6, 1985.
13. Azhermachev, A.K., Remnev, Yu.V., Chuprunov, V.P., Kolombet, L.V., Chugunov, V.A., Use of ultrafiltration in technology for isolating avian myeloblastosis virus, *Biotekhnologiya*, No.1, 1989 (cover to cover translation).
14. Ibid.
15. Official website of the Federal Medical-Biological Agency's Federal State Unitary Enterprise "State Scientific-Research Institute of Highly Pure Biopreparations", https://hpb.spb.ru/ru/institute/history/, Accessed on 20 August 2020.
16. Bykov, V.A., Ob utverzhdenii Polozheniya o Glavnom upravlenii "Biopreparat" pri Ministerstvo meditsinskoi i mikrobiologicheskoi promyshlennosti, Prikaz, No. 351, 23 April 1986, *RGAE*, Fond 788, Opis' 1, Delo 28, pp. 194–210.
17. Hoffman, D.E., *The Dead Hand*, p. 332.

18. Istoricheskaya spravka, *RGAE*, Fond 788, Opis' 1, p. 1.
19. Spravka ob itogakh raboty Nauchno-tekhnicheskogo soveta za 1986 god i proekte plana raboty na 1987 god, *RGAE*, Fond 788, Opis' 1, Delo 326, pp. 1–7.
20. Domaradskij, I., Orent, W., *Biowarrior*, p. 437.
21. Mangold, T., Goldberg, J., *Plague Wars*, p. 99.
22. Domaradskij, I., Orent, W., *Biowarrior*, p. 190.
23. Mangold, T., Goldberg, J., *Plague Wars*, p. 99.
24. Istoricheskaya spravka, *RGAE*, Fond 788, Opis' 1, p. 5.
25. Ibid., pp. 5–6.
26. *Materialy vsesoyuznogo simpoziuma posvyashchennogo 60-letiyu Tbilisskogo NIIVS*, Tbilisi, 1984, p. 13.
27. Guillemin, J., *Anthrax: The Investigation of a Deadly Outbreak*, University of California Press, Berkeley, 1999, pp. 133–134.
28. Information provided by Senior Scientific Worker, Immunobiological Laboratory, Eliava Institute of Bacteriophage, Microbiology & Virology, 30 October 2000. Guillemin appears to suggest in her account that the *STI* vaccine which was used in Sverdlovsk in 1979 had in fact been stockpiled by the authorities for more than a decade and was actually produced by the Tbilisi institute on 12 June 1964. Guillemin, J., *Anthrax*, p. 134.
29. Bykov, V.A., On the creation of a branch system of breeding centres for laboratory animals, *Minmedbioprom* Order No. 491, 28 July 1988, *RGAE*, Fond 788, Opis' 1, Delo 850, pp. 182–183.
30. Staff List of the Central Apparatus of the Ministry of the Medical Industry for 1990, 12 November 1990, *RGAE*, Fond 788, Opis' 1, Delo 1395, p. 7.
31. Ibid., p. 7.
32. Alibek with Handelman, *Biohazard*, p. 185.
33. Alibekov, K.B., Prilozhenie 1 k prikazu Ministerstva meditsinskoi promyshlennosti SSSR ot 22.08.90 No. 299, *RGAE*, Fond 788, Opis 1, Delo 1345, pp. 185–186.
34. Nazarenko, A.V., Sostav ekspertnoi gruppy dorabotke proekta kontseptsii razvitiya meditsinskoi i mikrobiologicheskoi promyshlennosti do 2005 goda, Prilozhenie k prikazu Minmedbioproma SSSR, No. 268, 16 June 1989, *RGAE*, Fond 788, Opis 1, Delo 1170, pp. 129–130.
35. There appears to have been serious problems with regard to quality-control in the production of these diagnostics. In 1988, Aleksandr Ivanovich Kondrusev, a USSR Deputy Minister of Health, offered the following complaints about the HIV test kits being turned out by *Vektor*: "*…poor quality was the reason why the USSR Ministry of Health recently rejected some test systems produced by the USSR Ministry of the Medical and Microbiological Industry's Vektor Science Production Association. This is all the more annoying since the test system employed a foreign-made chemical reagent that we*

purchased in limited quantities". See Rimmington, A., Oxford Virology tap Soviet biotech expertise in ambitious new venture, *The Genetic Engineer and Biotechnologist*, Vol. 11, No. 1, January/February 1991, p. 5.

36. Bykov, V.A., O podgotovke novykh vaktsinnykh i diagnosticheskikh shtammov virusa grippa i ikh vnedrenii v proizvodstvo vaktsin i diagnosticheskikh preparatov, Prikaz, No. 150/140, 11/13 April 1990, *RGAE*, Fond 788, Opis 1, Delo 1341, pp. 83–85.
37. Alibekov, K.I., Prilozhenie No. 1 k prikazu Ministerstva meditsinskoi promyshlennosti SSSR, No. 122, 5 April 1990, *RGAE*, Fond 788, Opis 1, Delo 1341, pp. 9–10.
38. *RGAE*, Fond 788, Opis' 1, Delo 1379, Report of Scientific-Technical Council for Work in 1990, 21 January 1991, 3 pp.
39. *RGAE*, Fond 788, Opis' 1, Delo 1377 Decisions Nos. 1–6 of the Scientific-Technical Council (NTS) Bureau and Documents to Support Them, 21 March to 30 August 1990, pp. 183.
40. Hodgson, J., Vector may help direct Oxford Virology, *Bio/Technology*, Vol. 9, February 1991, p. 125.
41. The Soviet delegation were invited to the UK by a small innovative biotechnology company, GB Biotechnology, based in Swansea, Wales. Among the scientists in the group were Academician Evgenii Davidovich Sverdlov and Academician Dmitrii Georgievich Knorre. The author assisted GB Biotechnology with the organization of the visit which included a memorable afternoon on the beach at Rhossili Bay on the Gower Peninsula. After their swim, the Soviet scientists were served with gin and tonics—the gin distilled under license by GB Biotechnology's director, Rod Greenshields. On 20–21 July 1989, the Soviet delegation participated in a biotechnology conference, *Biotechnology for the 21st Century*, held at St Catherine's College, University of Oxford. It was at this event that initial contact was made by Soviet scientists with representatives of Oxford Virology plc. See Fishlock, D., Help-line to Moscow, *Financial Times*, 14 October 1989; and Fishlock, D., Biotech windfall from the Soviets, *Financial Times*, 3 January 1991.
42. Ibid.
43. Personal communication to author from Derek Lennon, Chairman of Oxford Virology plc, DCL/3540R.rek, 18 October 1990, pp. 2.
44. Ibid.
45. Highfield, R., Soviet drugs could be used to treat UK patients, *Daily Telegraph*, 21 May 1991, p. 4; *NPO Vector, Novosibirsk, Siberia, USSR, Biological Sciences—Research and Production, Products*, Distributed by Oxford Virology plc, London, 75 pp.
46. Hodgson, J., *Bio/Technology*, Vol. 9, February 1991, p. 125.
47. Telex from Alibekov to Mr Derek Lennon, 12:46, 5 December 1990.

48. Personal communication to author from Derek Lennon, Chairman of Oxford Virology plc, DCL/3944R.rek, 14 December 1990, pp. 1.
49. Ibid.
50. Rimmington with Greenshields, *Technology and Transition*, p. 98.
51. Mityushev, *Zapiski obyknovennogo cheloveka*, p. 487; Alibek with Handelman, *Biohazard*, p. 194.
52. Hoffman, D.E., *The Dead Hand*, p. 355.
53. Alibek with Handelman, *Biohazard*, p. 196.
54. Hoffman, D.E., *The Dead Hand*, pp. 353–354.
55. Fenner, F., Henderson, D.A., Arita, I., Jezek, J. and Ladnyi, L.D., *Smallpox and Its Eradication*, World Health Organization, Geneva, 1988, p. 1431, https://apps.who.int/iris/bitstream/handle/10665/39485/9241561106.pdf?sequence=1&isAllowed=y, Accessed on 8 October 2020.
56. Kelly, D.C., The Trilateral Agreement: Lessons for Biological Weapons Verification, p. 95.
57. Ibid.
58. Alibek with Handelman, *Biohazard*, pp. 205–206.
59. Ibid., p. 242.
60. Ibid., pp. 257–258, 267.
61. Ibid., pp. 184–185.
62. Ibid., p. 184.
63. Leningrad, 0530 GMT, 3 June 1987, translated in *Summary of World Broadcasts*, SU/W1449/A/10, 10 July 1987, p. A10.
64. Dolot, E., Plant closed as problem is aired, *Soviet Union*, No. 1(454), 1988, pp. 48–51.
65. Rimmington, A., Biotechnology falls foul of the environment in the USSR, *Bio/Technology*, Vol. 7, No. 8, pp. 786–787.
66. Walker, M., Bio-plant poisons town, *The Guardian*, 16 March 1988.
67. Sidorov, I., Vozdukh dolzhen byt' chistym!, *Leningradskaya Pravda*, No. 127(21,969), 3 June 1987.
68. Rimmington with Greenshields, *Technology and Transition*, p. 49.
69. Faktuldinova, I., Some secrets from the secret "Progress": We are not making biological weapons, *Leninskaya smena*, 4 April 1991, p. 2, translated in *JPRS Report, Science & Technology, Central Eurasia: Life Sciences*, JPRS-ULS-91-016, 10 September 1991, p. 32.
70. Alibek with Handelman, *Biohazard*, p. 117.
71. Fedorov, L., *Soviet Biological Weapons: History. Ecology. Politics*, pp. 127–128.
72. Istoricheskaya spravka, *RGAE*, Fond 788, Opis' 1, pp. 9–10.
73. Ibid., pp. 11–12.

CHAPTER 8

A Brave New World: Building Capitalism in the New Russia and the Struggle for Control of *Biopreparat*

The collapse of the Soviet Union took place with dramatic speed in the last quarter of 1991. On 26 December 1991, the USSR Supreme Soviet abolished itself and declared that "*the Soviet Union no longer exists*".[1] Three weeks earlier on 5 December 1991, the State Concern *Biopreparat* was transferred to the newly created Russian Ministry of Health. The latter was itself reorganized in January 1994 as the Ministry of Health and the Medical Industry of the Russian Federation—*Ministerstvo zdravookhraneniya i meditsinskoi promyshlennosti Rossiiskoi Federatsii* or *Minzdravmedprom*.[2] The new minister was Colonel-General (Reserves) Eduard Aleksandrovich Nechaev, who had previously served as chief of the USSR Ministry of Defence's Main Military-Medical Administration.[3] Nechaev's ministry was now by far the most significant organization with regard to the management of applied and industrial biotechnology in the Russian Federation. Within its vast network were included 32 research institutes and colleges, 101 immunobiological and pharmaceutical production facilities, 29 facilities producing various medical equipment, 21 farms and most significantly in the context of this report, 60 "*conversion enterprises*" (which presumably encompassed the *Biopreparat* network).[4] Nechaev, who remained as minister until late November 1995, expressed himself determined to utilize Russia's own production capacities in the pharmaceutical and microbiological industries to eliminate shortages of key preparations. He drew up a list of pharmaceuticals which should be

A. Rimmington, *The Soviet Union's Invisible Weapons of Mass Destruction*, https://doi.org/10.1007/978-3-030-82882-0_8

put into production within the following two years, including insulin, anaesthetics, hormone preparations and broad-spectrum antibiotics.[5]

In May 1994, the Russian Joint Stock Company *Biopreparat* (*Rossiiskoe aktsionernoe obshchestvo Biopreparat* or *RAO Biopreparat*) was created on the basis of the State Concern *Biopreparat*.[6] Under the terms of the legislation, the state retained 51 per cent of the authorized capital.[7] No information has been released regarding the recipient(s) of the remaining 49 per cent of the equity and *Biopreparat* officials declined to identify any of its major shareholders. There were no major upheavals with regard to the control of *Biopreparat* at this time. Yurii Kalinin retained his position as chairman and was appointed to the Board of Directors by Order No. 1315-r of the government of the Russian Federation "Concerning the Appointment of State Representatives to the Council of Directors of *RAO Biopreparat*" (*O naznachenii predstavitelei gosudarstva v sovet direktorov RAO "Biopreparat"*) which was issued on 12 August 1994.[8] Among the other appointees to the board in 1994 was Lieutenant-General (Reserves) Valentin Ivanovich Evstigneev, the former head of the USSR Ministry of Defence's Fifteenth Directorate with overall control of the Soviet offensive biological weapons programme. At the time of the decree in 1994, he had been serving as deputy head of the Russian defence ministry's Radiation, Chemical and Biological Defence Forces (*Voisko radiatsionnoi, khimicheskoi i biologicheskoi zashchity, RKhBZ*).[9] Evstigneev's position on the board was renewed on 8 May 1999.[10]

There is evidence that the Russian Ministry of Health and the Medical Industry was subsequently unable to stamp its authority over *Biopreparat*. The situation as of 1995 was summed up by Yurii Fedorovich Doshchitsyn, then head of *Minzdravmedprom*'s Administration of the Medical Industry (*Upravleniya medpromyshlennosti*), who was the official with direct responsibility for the organization. He reported that his Administration had "*been unable to sort out its relations with Biopreparat. The reason for this is that Biopreparat has recently succeeded in securing budget finance outside of Minzdravmedprom and, having acquired economic independence, is attempting to take factories of the same profile with it*".[11] The identity of the organization responsible for funding *Biopreparat* at this time is unknown.

Kalinin's relations with the Russian Ministry of Health are then reported to have severely deteriorated. In addition, there were a series of reports in the Russian media voicing criticisms of Kalinin culminating in an extremely hostile attack in *Komsomol'skaya pravda* on 7 December 2000.[12] In April 2001, at an extraordinary meeting of shareholders of *RAO Biopreparat*,

Kalinin lost the support of the state, represented by the Ministry of Property (*Minimushchestvo*), and of private investors. In the ensuing election to determine the leadership of the joint stock company, Kalinin was defeated by Dr Ramil Usmanovich Khabriev, a civilian public health expert, by a margin of 73 per cent of votes to 27 per cent. Khabriev had previously been Minister of Health of the Republic of Tatarstan and then head of the Russian Ministry of Health's Department for Control of the Quality of Pharmaceuticals and Medical Technology.[13] The appointment of Khabriev from within the Russian Ministry of Health appeared to indicate that the ministry would now have very much stronger influence in decision-making within *RAO Biopreparat*.

Upon his appointment Khabriev asserted that his first action as general director would be to restore control of *RAO Biopreparat* over Russian antibiotics plants controlled by the Joint Stock Company "*Sintez*" (Kurgan), the Joint Stock Company "*Biosintez*" (Penza) and the Joint Stock Company "*Biokhimik*" (Saransk). Apparently, Kalinin had previously ordered that these facilities be transferred to the "*Farmtek*" group which was at that time controlled by *RAO Gazprom*. Khabriev argued that *Biopreparat* should have "*...restored the government's influence over the large plants either by purchasing or exchanging additional blocks [of shares.... It is now virtually impossible to buy shares of these plants—the time has passed....But there are compromises. An agreement can be reached with shareholders or the plants themselves to assure them that they can work long and productively with the Russian joint-stock company*".[14] Khabriev floated the possibility that he would create a subsidiary company under *RAO Biopreparat* with direct responsibility for the management of these plants. He does not, however, appear to have been successful in his attempt to recover ownership of the antibiotics' enterprises or other large facilities.

Kalinin's Retention of Power Within Russia's Pharmaceutical Sector

Despite his dismissal as head of *RAO Biopreparat*, Kalinin continued to hold key levers of power within the Russian pharmaceutical sector. Most importantly, he continued in his position as President of the Russian Association of Producers and Suppliers of Pharmaceuticals, Medical Products and Technologies (*Rossiiskaya assotsiatsiya proizvoditelei i postavshchikov lekarstvennykh sredstv, izdelii i tekhniki meditsinskogo naznacheniya* or *Rosmedprom* for short).[15] This organization played a pivotal role within the Russian pharmaceutical industry, incorporating more than 70 enterprises and organizations, and acting as the industry's principal channel of communication with government departments.[16]

Even before his dismissal as head of *Biopreparat* however, Kalinin's management of *Rosmedprom* was being criticized by those wishing to pursue wholesale reform of the sector. In February 2000 for example, the head of the Russian Ministry of Economics' Department of Economics of the Medical Industry, A. Tereshchenko, criticized Kalinin and M.I. Grigor'ev (Deputy General Director of *RAO Biopreparat*) for their management of the *Rosmedprom* Association. He noted that *Rosmedprom* has played an important role in implementing positive changes within the medical industry including coordination of the activities of enterprises and participation in the elaboration of measures of state support for Russian producers. "*Undoubtedly the contribution of Kalinin and Grigor'ev, who are high-quality specialists in their sphere, has been significant.*"[17] However, in Tereshchenko's opinion in the changed market conditions of the day, the Association of Russian Producers should be led by somebody whose work in the Association was their main task. This he believed would facilitate the improvement of the role of the Association in the resolution of the strategic questions facing the pharmaceutical industry in Russia. In what can only be viewed as a veiled threat, Tereshchenko warned Kalinin that the directors of Russian pharmaceutical enterprises might consider establishing a new Association if he did not hand over power to a new management team.[18] Despite these criticisms, Kalinin continued in his position as President of the Association.

Kalinin continued also to be appointed as a member of key government committees including the Russian Federation Security Council Interdepartmental Health Protection Commission and the Interdepartmental Commission on Problems in Genetic Engineering Activity.[19] Kalinin was also at this time Chairman of the NASA-funded "*Space Biotechnology*" grant programme committee.[20]

The Establishment of the Open Joint Stock Company *Biopreparat-Tsentr* (*Biopreparat*-Centre)

In 1993, a company which interacted extremely closely with *RAO Biopreparat* is reported to have been created by Yurii Kalinin. It was given the confusingly similar name Open Joint Stock Company *Biopreparat-Tsentr* (*Aktsionernoe obshchestvo otkrytogo tipa Biopreparat-Tsentr* or *OAO Biopreparat-Tsentr*)—shares in an *OAO* can be traded publicly—with Kalinin taking the position of director. Shortly prior to his dismissal as director of *RAO Biopreparat*, Kalinin is alleged to have transferred the ownership of the buildings at Samokatnaya ulitsa to *OAO Biopreparat-Tsentr*.[21] From this point onward, the two organizations then shared the same headquarters building and Yurii Kalinin is reported to have retained his former director's office.[22] This was not a temporary transitional occupation since, as late as 2002, a visitor to *Biopreparat*, observed that Kalinin occupied an office directly opposite to that of the then director of the company, Ramil Usmanovich Khabriev. Indeed, Khabriev was in the humiliating position of having to rent his office at the site from Kalinin.[23]

One of the prime motivations behind the creation of the new spin-off company appears to have been to gain control of commercially valuable technologies and property belonging to *RAO "Biopreparat"* and its subordinate facilities. A report in *Rossisskaya gazeta* in December 2001 outlined how *OAO Biopreparat-Tsentr* had, in November 1999, become the owner of an experimental-industrial process for production of recombinant human insulin. This technology and the recombinant strain had originally been developed in 1999 by the State Scientific Centre of Applied Microbiology (Obolensk) although the patent holder was *RAO Biopreparat*. As the newspaper noted, "*...both the state, which funded the work from the outset, and the scientists who invested much effort and grey matter in the creation of the said technologies, were left holding the bag—quietly and unnoticed a private structure had become the de facto owner of unique technologies which had been developed over many years at a state scientific research institute and with state money to boot*".[24]

There are some indications that there may subsequently have been a reduced state role for *Biopreparat-Tsentr*. In 2004, for example, it was removed from the list of Russian strategic facilities. This was then followed by a Russian government announcement in August 2005 that, the following year, it planned to privatize 10 per cent of *Biopreparat-Tsentr* along with other companies.[25] However, it is notable that at the Inter-State

Scientific-Practical Conference "*Emergency Situations of International Significance in Public Health & the Sanitary Protection of the CIS States*" scheduled for Obolensk in October 2006, it was envisaged that two representatives of *Biopreparat-Tsentr* would be invited to participate in this prestigious event.[26]

At the time of writing, Kalinin was celebrating his 82nd birthday. Neither his state of health nor the extent to which he remains active in the Russian biopharmaceutical sector is known. However, the likelihood is, that his current involvement has been considerably reduced and he may now be holding honorific positions rather than wielding real power and control. Clearly, his role, with regard to both the management of the *Biopreparat* network and its subsequent breaking up and redirection in the Russian Federation, has been enormously significant. He remains a highly controversial figure, with both critics and plaudits, and would clearly be a fascinating subject for a biography if a Russian writer brave enough would come forward.

Proposed *Biopreparat* Role in the Conversion of Russia's Military Biological Facilities

There is evidence that, besides its key role in *Minzdravmedprom*'s plans for pharmaceutical production, *Biopreparat* was also being lined up to manage the conversion of biological defence facilities belonging to the Russian Ministry of Defence. In February 1992, Boris Yeltsin's government created a new organization, the Committee on Convention Problems of Chemical and Biological Weapons (*Komitet po konventsial'nym problemam khimicheskogo i bakteriologicheskogo oruzhiya*).[27] It was entrusted with the responsibility for the formulation and implementation of a coordinated conversion programme for the Ministry of Defence's facilities. The Committee was headed by Pavel Pavlovich Syutkin who had recently replaced Lieutenant General (Reserves) Anatolii Dem'yanovich Kuntsevich.[28] It operated on a budget of around US$290,000 and was sub-divided into three departments concerned with destruction of chemical weapons, ensuring compliance with chemical weapons agreements and ensuring compliance with the 1972 Biological and Toxin Weapons Convention.[29]

The Committee was exercising close control over the conversion of military microbiological facilities. According to a 1992 report in the newspaper *Izvestiya*, experts from the Committee held the view at this time that the defence ministry's military biological facilities based at Kirov, Sergiev Posad and Ekaterinburg should, over time, be removed from military jurisdiction and switched to work for the civil economy. In order to achieve this, they put forward the idea that a single science production complex (*nauchno-proizvodstvennyi kompleks*) should be created for the management of coordinated R&D and production at these centres.[30] It was against this background, that at some point between 1992 and 1994, a representative from the US investment bank and stock brokerage firm PaineWebber Incorporated, held a meeting with Kuntsevich and senior colleagues which was specifically focused on the potential for cooperation in infectious diseases in animals and production of veterinary vaccines with Compound No. 19 (Ekaterinburg). The Committee's chief engineer, General A. Blatov, who had been given specific responsibility for Compound No. 19 was in attendance. However, the key participant in the meeting appears to have been Anatolii D. Smirnov from *Biopreparat*. It was proposed that the latter would act as a holding company for any potential ventures which were created.[31] Although the project eventually floundered because of the military's desire to maintain the "closed" (highly restricted access) status of its biological facilities, it does highlight the central role that *Biopreparat* continued to play in decision-making and management of Russia's military biological complex at this time.

The Financial Meltdown Within *RAO Biopreparat* and the Pivotal Role of the ISTC

During the 1990s and early 2000s, the network of R&D facilities controlled by *RAO Biopreparat* faced an acute funding crisis. Despite the fact that Kalinin, before his recent dismissal, sold parcels of shares in several of his research institutes to *Gazprom* and other financial interests, none of this revenue was invested in scientific research programmes or in the maintenance of facilities. The Institute of Immunological Engineering—*III* (Lyubuchany) provides a good example of the financial problems faced by the *RAO Biopreparat* research facilities at this time. From 1995 to 2001, Kalinin had apparently been unable to provide any funding for this institute at all. Only 13 per cent of the institute's budget was provided by the state, via the Russian Ministry of Health and the Russian Ministry of Industry, Science and Technology. The remaining 87 per cent of the budget was in fact being provided by the International Science and Technology Centre (ISTC)—an intergovernmental organization established in 1992 by an agreement between the European Union, the US, Japan and the Russian Federation. Funded by the US and its allies, the ISTC was originally focused on engaging former weapons scientists and redirecting their expertise to non-military research. The financial crisis was so acute at *III* that the institute faced imminent bankruptcy, and by 2001, its burden of debt stood at approximately 50 per cent of its total estimated economic value.[32]

A similar situation confronted the State Scientific Centre of Applied Microbiology in Obolensk. The operating budget of the facility had fallen precipitously from approximately US$25 million in 1991 to around US$2.5 million in 1999.[33] The main reason for the Centre's penury appears to be that its most profitable units, such as its nutrient media plant and recombinant insulin production, were privatized and transferred by *Biopreparat* to a range of private companies and investors. This left the facility with no substantive revenue-generating capability of its own. By 2000, divested of its revenue-generating operations, the applied microbiology facility was in serious financial difficulties and reported to be on the brink of shutdown.

The dependency on international funding had never been greater and Smithson reports that in 1999 the Obolensk facility was receiving roughly as much financial support from the ISTC as from the Russian government and at any one time, roughly 50 per cent of its staff relied on ISTC grants.[34] The situation was so grave that on 18 September 2000, the Centre's Director Nikolai Nikolaevich Urakov addressed the following letter to Gennadii Nikolaevich Seleznev, Chairman of the Russian Federation State Duma: "*Dear Gennadii Nikolaevich! In the five months since our previous communication to you, it has not been possible to stabilize the economic situation of the State Scientific Centre for Applied Microbiology....The apparent conclusion is that either the sovereign authority does not have objective information about the tragedy of applied science in Russia or else it is intentionally following a course of the sector's total destruction...*".[35]

The situation at Obolensk went from bad to worse and in March 2002, *Moscow Times* reported that the Moscow utility, Mosenergo, was threatening to cut electricity to the Obolensk facility because it had not paid its bill in more than three years and owed US$1.4 million.[36] In April 2002, both the Mosenergo and Mosregiongaz utilities took the facility to court for recovery of a combined debt of US$2.5 million. Later, in October 2002, it was reported that a Russian judge had issued a ruling permitting bankruptcy proceedings against the Centre because of its unpaid energy bills. Urakov, Director at Obolensk, had argued that his facility simply lacked the means to pay this debt, not having received any federal funds for the past five years.[37] As late as the winter and spring of 2005, water supplies to the settlement of Obolensk were threatened with restriction because of the non-payment of 38 million roubles accumulated debt by the institute.[38] In June of that same year, collective wage arrears at the institute stood at Roubles 16.5 million.[39]

By 2000, the State Scientific Centre of Virology and Biotechnology *Vektor* (Kol'tsovo, Novosibirsk *oblast'*) was also suffering a massive contraction of its operational budget, and because of indebtedness to its energy suppliers received only an emergency-reserve level of power.[40] At some point in 2000, Professor Lev Stepanovich Sandakhchiev, Director of "*Vektor*", directed a letter to N.F. Gerasimenko, Chairman of the Russian State Duma Committee on Health Protection. He noted that his facility "*has become indebted to energy suppliers, and the research centre is receiving only an emergency-reserve level of power. As a result, work with ultra-dangerous infections has been halted...*".[41] In a desperate effort to identify new sources of funding, Sandakhchiev later unveiled a US$25 million proposal to transform his facilities into an International Centre for the Study of Emerging and Reemerging Infectious Diseases (ICERID).[42] In September 2005, Sandakhchiev was honoured by the US Civilian Research and Development Foundation in recognition for his generous support to the CRDF in developing international collaborations in science and technology between the US and Russia.[43] Upon his death in 2006 (see below), *The Washington Post* noted that with major financial and technical backing from the US, Sandakhchiev had eventually presided over "*Vektor's conversion into one of Russia's leading centres for disease surveillance and research*".[44]

Meanwhile, *RAO Biopreparat* and the military were all too aware that concerns over the global proliferation of weapons of mass destruction would force the US and its allies to intervene to provide substitute funding for Russia's debt-ridden former BW institutes. The *RAO Biopreparat* leadership perceived this international funding as being a useful means of allowing its institutes to remain intact and operational in the absence of alternative domestic financial resources. Kalinin however, wished to keep close control over outside engagement with his facilities. In April 1997, for example, Kalinin is reported to have met a US-IPP delegation and refused their requests to visit Obolensk, Lyubuchany and *NIIbiokhimmashproekt* (Moscow) despite the fact that the delegation had letters of invitation from the directors of all three establishments.[45] However, in the very same month, he appeared much more receptive to an initiative made by the National Academy of Sciences' Committee on US-Russian Cooperation on Dangerous Pathogens. Here, he played a central role, leading the Russian delegation and hosting a meeting with US representatives in the Moscow suburb of Petrovo-Dalnyee during 28–29 April 1997 to consider the general framework for cooperation. Discussions subsequently continued at a smaller follow-up meeting organized by the NAS Committee in Moscow in June 1997, attended by representatives of *Biopreparat* and directors of several of its key research institutes.[46] In September 1997, Kalinin, also had a meeting at his headquarters building with Andy Webber of the Department of Defence's Office of Threat Reduction and a small delegation of US officials. Kalinin was again emphatic that all US assistance should be channelled through *Biopreparat* rather than via direct relationships with the institute's themselves.[47]

Another manifestation of *Biopreparat*'s closer relations with US entities at this time concerns a joint project which it entered into with G.D. Searle, which at that time was part of the Monsanto group. In 1997, the US company invested in a new US$32 million joint venture with *Biopreparat*'s All-Russian Centre for Molecular Diagnostics and Therapy. Searle Pharma and *Biopreparat* owned the business jointly, having 75 per cent and 25 per cent shares respectively.[48] On 15 September 1999, the Searle-*RAO Biopreparat* joint venture began operation of its pharmaceutical production facility at Izvarino (Moscow *oblast'*).[49] The Izvarino site was the location of several incomplete P4 containment facilities which had been scheduled to be built for the Soviet-era All-Union Scientific-Research Institute of Applied Molecular Biology, a facility originally under the control of the Third Main Administration of the USSR Ministry of Health.[50] The site was also intended to incorporate industrial production capacity and the foundations had previously been laid for a manufacturing plant. The new Searle Pharma-*Biopreparat* plant came on stream in 2000 with an estimated capacity of 300 million tablets and capsules. The new plant was the first enterprise in Russia which was especially designed and constructed in full accordance with worldwide GMP standards, which guaranteed the safety of its production. It was primarily focused on the manufacture of new generation cardio-vascular, gastroenteric and antibacterial drugs.[51] In many ways, the year 1997 with the issuance of a raft of joint initiatives and projects can be viewed as representing the high watermark in relations between *Biopreparat* and the West.

The Fight for Ownership of *Biopreparat*'s Assets

One of the earliest conflicts surrounding ownership of *Biopreparat*'s assets concerns one of the joint stock company's most important facilities, the Science Production Association *Biomash* (*NPO Biomash*). As detailed above, the *NPO* was headed by the Moscow-based Scientific-Research Design Institute of Applied Biochemistry (*Nauchno-issledovatel'skii institut prikladnoi biokhimii* or *NIIbiokhimmashproekt*). During the Soviet period, *Biomash* had been the nerve centre of *Biopreparat* and was responsible for the design and fitting-out of facilities across the Soviet Union. In January 1993, a 1000 m^2 incomplete production facility called Building D, which originally belonged to *Biomash* and is located on ulitsa Klary Tsetkin, was seized by a group of men wearing paratrooper uniforms and armed with police truncheons. They were acting on behalf of the Joint Stock Company *Apromed* which claimed the building—valued at Roubles 1.1 million—had been legally privatized and belonged to them. Despite the fact that the State Concern *Biopreparat* declared *Apromed*'s original acquisition of this facility to be illegal, the Joint-Stock Company planned to demolish part of the fence surrounding *NPO Biomash* and create a gate allowing access to Building D.[52] *Apromed* was in fact established by two deputy directors of *NPO "Biomash"* who appeared to be seeking to acquire part of this *Biopreparat* facility for their new company.[53] Despite opposition from Alibekov, who was the *Biomash* director at this time, P.P. Mostov, the deputy head of the State Property Committee subsequently issued documentation confirming the legality of the transfer of Building D to *Apromed* which continues to operate today at ulitsa Klary Tsetkin.[54] It is not known what role, if any, Kalinin, behind the scenes, may have played in the transfer in ownership of this facility.

The incident involving *Apromed* provided an extremely rare public glimpse of the struggles at this time, involving *Biopreparat* assets, which in all likelihood, were widespread but invisible to outside view. Another incident which made its way into the Russian media took place on 13 July 2005 and involved Sergei Anisimov, the then general director of the Science Production Association *Elevar* (*NPO Elevar*). At this time, *NPO Elevar* was a construction-oriented company "*that has very profitably built food processing plants, breweries, and other facilities related to the strengths and interests of Biokhimmash*".[55] After leaving his office at the end of the working day, Anisimov was shot five times and wounded in his car on ulitsa Klary Tsetkin in Moscow. The assassination attempt was linked in the media at the time to an unresolved dispute with a sub-contracting company in Omsk where *Elevar* is reported to have been involved in a number of substantive projects at local enterprises.[56]

Biopreparat and the Bitter Struggle for Control of Lyubuchany

The best-documented case detailing the intense struggle for control of *Biopreparat* assets concerns events surrounding the Institute of Engineering Immunology (*Institut inzhenernoi immunologii* or *III*)—also known as Post Office Box *G-4883*—in Lyubuchany (Chekhov *raion*, Moscow *oblast'*). Before examining these in any detail, it is worth briefly reviewing the history of this important *Biopreparat* institute. The facility was created, originally as the Institute of Immunology, on 16 November 1978 by Decree No. 920 of the USSR Council of Ministers. Under the terms of this document, the institute was to undertake priority tasks in the fields of immunology, microbiology, bioengineering and biotechnology at the order of the USSR Ministry of Defence.[57] Leitenberg and Zilinskas report that there was a second secret section of the decree which pointed to the real mission of *III*. This embraced an assessment of "*the immune response of animals to pathogens of BW agents, to discover immune system weaknesses that could be exploited by new BW agents, to overcome immune response induced by current vaccines, and to develop vaccines to protect Biopreparat scientific workers from the pathogens on which they worked*".[58]

The institute, which was eventually established on a 13.2-hectare site at Lyubuchany in 1979, was in fact an offshoot of the All-Union Scientific-Research Institute of Applied Microbiology's (*VNIIPM*'s) Laboratory of Molecular Biology headed by Vladimir Petrovich Zav'yalov (see Fig. 8.1).[59] The first director of the new spin-off was Colonel (Reserves) Vladimir Davidovich Savve, a graduate of the S.M. Kirov Military-Medical Academy (Leningrad), who had formerly been employed within the USSR Ministry of Defence's Sanitary Institute in Zagorsk. Here he had been engaged in work on the detection of variola virus and the early diagnosis of smallpox. Savve's appointment as director was supported by Vorob'ev, an

Fig. 8.1 Professor Vladimir Petrovich Zav'yalov, Director, *Biopreparat*'s Institute of Immunology, May 1994 (©A. Rimmington)

ex-colleague at the Sanitary Institute, despite Popov's support of Zav'yalov for assuming this position.

Initially at least, *III* was housed within a building belonging to *VNIIPM* in Obolensk. Here, in 1980, Zav'yalov took up the position of head of the Laboratory of Molecular Immunology. A year later he was appointed the head of *III*'s Department of Molecular Immunology.[60] It is likely that Zav'yalov's civilian background, including his work up until 1977 at the Crimean Medical Institute's Department of Biochemistry, initially hindered his appointment to the most senior positions at the *Biopreparat* facility. A search for new accommodation was initiated. At first the institute was relocated to Dashkovka, near Bol'shevik, on the south western edge of Serpukhov. Eventually pre-existing facilities at Lyubuchany, around 37.5 km north-east of Obolensk, were selected. Here, *Glavmikrobioprom* had been constructing facilities, including a two-storey building housing a fermentation plant and vivarium, and a three-storey building housing laboratories, for a branch of its subordinate *VNIIgenetika* (see above). These buildings were now transferred to *III*. By 1984, a purpose-built administration block had been completed and was operational at the Lyubuchany site, with two buildings containing laboratories then being occupied a year later. Set against a background of trees, the institute was approached across a flat featureless plain making it the most visible landmark from the road. The entire site was protected by a security wall.[61]

In 1984, Zav'yalov joined the CPSU and was perceived as being more politically reliable and this allowed Savve to appoint him as his deputy.[62] In 1985, there arose the first in a series of conflicts which were to be a recurring feature of the relations between the senior management of *III* and that of *Biopreparat*. A conflict over their respective military ranking is reported to have led Kalinin to dismiss Savve in 1985. Zav'yalov immediately took up the role of acting director and in December 1986, this position was made permanent. The Ministry of Defence had apparently attempted to argue the merits of appointing their own candidate, Lieutenant Colonel (Reserves) Stanislav Stepanovich Afanas'ev. However, the final say over the issue is credited to Academician Ovchinnikov and the members of the Interagency Scientific-Technical Council (*MNTS*)—see above—who determined that Zav'yalov was the most well-qualified scientist to lead a programme focused on immunology at *Biopreparat*. The military's interest in the facility was recognized with the appointment of Afanas'ev as Zav'yalov's deputy. Another Colonel, Yu.I. Morozov, also served on Zav'yalov's staff, together with Viktor Isaakovich Marchenko, who had won the Lenin Prize in 1984 for his work on biologically active peptides.[63]

The Lyubuchany institute was engaged in a number of projects for the Soviet military. One of the most significant concerned the development of a new associated vaccine against 18 pathogenic microorganisms including plague, anthrax, typhoid, cholera and Venezuelan Equine Encephalitis. In conjunction with the USSR Ministry of Defence's Institute of Military Medicine (Leningrad), two-year clinical trials of the new vaccine were undertaken. Tableted interferon was also produced for the Soviet military and tests monitoring the prophylaxis of viral diseases in naval personnel were conducted.[64] The institute is reported to have pioneered the use of clean room technology in the Soviet Union and a suite of clean rooms for producing vaccines and pharmaceuticals was constructed at the site.

The crucial importance of *III* within the *Biopreparat* network is pointed to by the analysis of Leitenberg and Zilinskas who argue that the institute was responsible for two of the five major accomplishments of the Soviet second-generation BW programme in bacteriology. The first of these involved the coating of *F. tularensis* cells with Protein A to allow them to bypass the immunodefence system in humans. The large-scale production of Protein A is reported to have been achieved at the Omutninsk Chemical Factory (see above). The second is reported to have involved the genetic engineering of an F1-minus strain of *Yersinia pestis* to alter the antigenic presentation of the pathogen's surface. This would have led to delays in the detection and identification of the pathogen in any country attacked with biological weapons armed with this strain.[65]

Biopreparat's Engagement with Russian and US Commercial Companies at Lyubuchany

A unique aspect of the situation at the Lyubuchany site concerns *III*'s close interaction with Russian and US companies. As detailed below, Alibekov was a key figure in *Biopreparat* responsible for identifying land at Lyubuchany which could be used by the Moscow-based *Kristall* (Crystal) distillery to build a plant for the manufacture of miniature plastic bottles. It was intended that the miniature bottles, containing vodka marketed under the Stolichnaya and Golden Ring brands, would be served on flights. The *Kristall* director then entered into negotiations with Zav'yalov about building on the territory belonging to *III*. The talks were successful and a joint venture called *OOO Biokristall* was created in 1990. Over a two-year period, a new manufacturing plant was built at the site.[66]

The *Kristall* project eventually appears to have been cancelled after a US company began supplying vodka miniatures to the airlines. However, in 1997, the US company Alcoa leased the plant which was renamed CSI Vostok Ltd. (see Fig. 8.2) and was reported to have initially signed a five-year agreement (with an option for a five-year extension) with "*Biopreparat*". Zav'yalov also signed an agreement with *Kristall*. An annual rent is reported to have been paid by Alcoa, and part of the funds received by Zav'yalov were transferred to *Kristall* for repayment of the initial investment at the site. In September 1997, CSI Vostok Ltd. initiated the manufacture of plastic closures for the packaging and consumer goods market.[67] The Lyubuchany site is reported to have been particularly attractive to the US company because it is 2–3 kilometres away from any habitation and is on a site that has been internationally certified as "*ecologically pure*". Some revenue also appears to have been generated for

Fig. 8.2 Sign at Lyubuchany directing visitors to Alcoa CSI Vostok (©A. Rimmington)

Biopreparat as a result of the agreement. The Alcoa presence is reported to have substantially improved the infrastructure at the Lyubuchany site including renovation of the security wall, roads and so on. *OOO Biokristall*, based at the Lyubuchany site, continued to manage the relationship with Alcoa. Fifty per cent of the equity of the company is reported to have belonged to the *Kristall* distillery and 50 per cent to the institute. Another aspect of Alcoa's operation at Lyubuchany concerns the large tax revenues its production plant generated for the local authority. The Chekhov District Council was a major beneficiary of the US company's presence at the site and the revenues it received enabled it to make massive improvements in the provision of local services and the pursuit of civic projects. The latter include the construction of a modern, well-equipped sports stadium for the local population.[68] It is a matter of pure speculation, but it does appear to be a possibility that the subsequent struggle for the control of *III* may have manifested itself with such intensity, largely owing to its joint ownership of the cash-generative *OOO Biokristall*.

The New Russia and the Struggle for the Control of Assets at Lyubuchany

The collapse of the USSR proved to be especially catastrophic for *III*. In 1992, *Biopreparat* is reported to have ceased direct funding and to have removed all secret documentation. There was also a dramatic decrease in other sources of government funding. The cessation of funding was mirrored by the departure of staff with numbers falling dramatically. There had been 112 researchers on the institute's staff in 1990 including 2 professors, 5 doctors and 46 Candidates of Science.[69] By May 1994, when the author visited the facility, the number of research personnel had been sharply reduced. The facility was generating some income at this time from its participation in joint work with the Russian Ministry of Defence's Centre of Virology on the development and production of new tableted vaccines. The institute was also testing new associative vaccines in Russian troops.[70]

During his visit in May 1994, the author noted that one of the interesting aspects of the Lyubuchany facility at that time was that it embraced the full cycle of basic and applied R&D, fermentation (via a small plant) and production of finished preparations (via its own tableting factory). A new production facility was also observed to be under construction sited just beyond the institute's outer security wall. Zav'yalov anticipated that the funds required to complete this project would be made available in the latter half of 1994. It was intended at this time to produce tableted recombinant cytokines in this facility.[71]

Zav'yalov retained his position as director when his facility was reorganized in 1993 as the Open Joint Stock Company "Institute of Engineering Immunology" (*Aktsionernoe obshchestvo otkrytogo tipa Institut inzhenernoi immunologii* or *OAO III*). Initially the state retained a 49 per cent holding in the new company. Few details are known about the other shareholders although, by February 2001, *Biopreparat* is known to have held 10 per cent of shares with an additional 15 per cent being held by its subordinate Closed Joint Stock Company *Medbiotekhprom* (*Zakrytnoe aktsionernoe obshchestvo Medbiotekhprom*). The Russian government at this time controlled 37 per cent of the Lyubuchany joint stock company with the remaining shares being held by the researchers themselves.[72]

A key aspect of the situation at Lyubuchany during the 1990s had been a deterioration in the personal relations between Kalinin and Zav'yalov. This situation came to a dramatic head in the first days of the new millennium. On 18 January 2000, Zav'yalov was driving to work at *III* accompanied by his wife. However, *Biopreparat* security personnel refused to allow him access to the Lyubuchany site via the sliding entrance gate. They informed him that Grigorii Yakovlevich Shcherbakov (head of the Scientific-Technical Administration of *OAO Biopreparat*), in his position as head of the board of directors of the Joint Stock Company "Institute of Engineering Immunology", had issued an order removing Zav'yalov as director of the facility. The order apparently declared that Shcherbakov himself was appointed as the new director.[73]

Zav'yalov considered this action to be completely illegal since it was his understanding that only an extraordinary meeting of shareholders could remove him from his position. Shcherbakov was not present on site at this moment and Zav'yalov managed to gain entrance to his office by claiming that he needed to collect his "*inside shoes*". Upon returning to the facility, Shcherbakov is reported to have been shocked to find Zav'yalov present on site and telephoned Obolensk to request that their head of security be sent over to resolve the situation. The situation was extremely tense with neither of the strong-willed individuals willing to concede defeat. Zav'yalov later took a call from a senior *Biopreparat* manager and refers to personal threats made against his family. Whether, if true, these were made just to put pressure on Zav'yalov or represented something more sinister will never be known, but he was certainly frightened enough to immediately leave the site. Shcherbakov subsequently remained in position as director of *OAO III* until April 2000 when he was replaced by Professor Sergei Yur'evich Pchelintsev.[74] Commenting on the events at Lyubuchany, *The Wall Street Journal* stated that this "*was a slap in the face to American and European donors who thought they were funding the institute's conversion to peaceful research under civilian control*".[75]

It appears significant that Zav'yalov was not dismissed from *OAO III*. His crucial scientific role at the institute meant that he was retained as a senior research scientist and a principal scientific advisor to the Joint Stock Company "Institute of Engineering Immunology". However, it was immediately apparent, both to his colleagues and visiting scientists, that Zav'yalov felt humiliated and embarrassed in his new, much-reduced role. However, his opportunity to seek alternative employment opportunities outside of Russia was curtailed by the fact that his overseas passport was retained by the *Biopreparat* authorities.[76] After his dismissal as director, there was some considerable sympathy for Zav'yalov among the scientific community, and colleagues in Moscow rallied to his aid and secured for him a number of positions which enabled him to continue to actively pursue his research interests. Later in 2000, for example, Valerii Alekseevich Bykov, former USSR Minister of the Medical and Microbiological Industry, secured for him an appointment as Professor in the Department of Biomedical Technologies at the Russian University of People's Friendship (Moscow). A year later, in 2001, he was appointed concurrently as the head of the Laboratory of Hybridoma Technology at the Russian Research Centre for Molecular Diagnostics and Therapy (Moscow).

The real reason behind the dramatic battle for control of the Lyubuchany joint stock company may in fact have had little to do with clashes of personality at all. Rather, it may very well have been connected with subsequent moves to dispose of the bulk of its intellectual and physical assets. For, in November 2001, the Closed Joint Stock Company *Biokad* acquired, "*for a modest sum*", the bulk of the facilities and research personnel based at *OAO III*. *Biokad* also acquired the institute's substantial portfolio of intellectual property, which as is detailed above, was developed via massive Soviet investment. Control of these assets gave access to sufficient threshold technologies "*to last ten years*". *Biokad* formed a subsidiary, the Centre of Immunological Engineering (*Tsentr Inzhenernoi Immunologii* or *TsII*), to manage its part of the Lyubuchany site. The new centre is engaged in research on the creation of new producers of genetically engineered preparations and biologically active substances. It is also developing genetically engineered cytokines, antibodies and other recombinant proteins and peptides, and taking these through to pilot-scale production. In October 2010, *TsII* was reported to employ 22 scientists including 1 Doctor of Science and 6 Candidates of Science.[77]

It appears to be significant that *OAO Biopreparat* should have sought to retain control of a small part of the Lyubuchany site. As has been outlined above, *OAO III* is credited with being responsible for two of the five major accomplishments of the Soviet second-generation BW programme in bacteriology. Therefore, it may have been considered to have some strategic importance by the authorities. Evidence for this is the fact that the Institute of Engineering Immunology appeared on the Russian government's list of strategic facilities which was published on 6 August 2004. According to the Russian government, the list concerns enterprises that have "*strategic significance for ensuring the defence and security of the state, (and) the protection of the morals, the health, rights and legal interests of Russian citizens*".[78] As *Biopreparat* facilities became increasingly focused on commercial opportunities in the civil sector, this status as a "strategic facility" may very well have been later removed by the authorities.

Despite the retention of control of part of the *OAO III* site, one former *Biopreparat* scientist considers that *Biokad* recruited on contract many of the leading Lyubuchany scientists to work at its new Centre of Immunology. According to him, *OAO III* at this time lost its best scientists and was reduced to a small core of experts. Those who elected to remain under *Biopreparat* control were apparently concerned at the potential lack of job security offered by *Biokad*. *OAO III* appeared to now have few assets or revenue-generating capabilities of its own. In addition, it was not receiving any of the revenue generated by the agreement with Alcoa which operates a plant producing plastic closures at Lyubuchany. *Biopreparat* was apparently attempting to commercialize the remaining research at the institute, but some outside observers were sceptical of their chances of success.[79]

The Creation of the Open Joint Stock Company *Biopreparat*

At some point around March 2003, *RAO Biopreparat* was renamed as the Open Joint Stock Company *Biopreparat* (*Aktsionernoe obshchestvo otkrytogo tipa Biopreparat* or *OAO Biopreparat*). The reasons for the change and what implications it had for the company have not been revealed. By April 2006, again for reasons unknown, *OAO Biopreparat* had evacuated its headquarters building located at Samokatnaya ulitsa.[80] It was transferred to the buildings belonging to the Institute of Applied Biochemistry and Engineering (*OAO Biokhimmash*) on ulitsa Klary Tsetkin. This was also the location of a number of bioscience companies it was closely linked to and which formed a key component of its remaining commercial operations. The reasons behind the apparent relocation of *OAO Biopreparat* remain a matter of speculation. There could be unknown security considerations which lie behind the move. It may have formed part of an attempt to disentangle operations from those of *Biopreparat-Tsentr*, managed by Yurii Kalinin, and which remained located at Samokatnaya ulitsa. There may be economic considerations and a desire to cease rental payments for the lease of the Samokatnaya ulitsa offices. It could even be the case that, given Moscow's inflated real estate values, there was a wish to dispose of the old HQ building and its associated land.

In 2006 the senior management of *OAO Biopreparat* continued to be dominated by Russian military and intelligence officials. The general director of *OAO Biopreparat* was Vladimir Nikolaevich Kolesnikov who was appointed to this position in 2004. Very little information is available in the open literature with regard to his background. He was relatively young, having been born in 1957. Prior to his appointment, during the period from 2002 to 2004, he served as a consultant in the government apparatus, although the precise position he held is unknown. In the past he had been a close associate of Yurii Kalinin, having served as a deputy general director of *RAO Biopreparat* from 1999 to 2002. He concomitantly served as the executive director of Kalinin's *Rosmedprom* Association.[81] The pivotal figure operating within *OAO Biopreparat* in 2006 appeared to be Lieutenant General (Reserves) Valentin Ivanovich Evstigneev who at this time held the post of first deputy director. He, more than any other member of the senior management, embodied the extremely close links which had previously existed between the company and the Russian Ministry of Defence. Evstigneev was appointed to this position on 2 January 2001 after the adoption by *Biopreparat* of a newly constituted charter which obliged the first deputy general director of the company to have the appropriate security clearance. He concomitantly served as the director of the company's Centre for Export Control.

Another important figure within *OAO Biopreparat* in 2006 was Colonel (Reserves) Grigorii Yakovlevich Shcherbakov, head of the Scientific-Technical Administration, who is reported to have joined the organization in 1989. Shcherbakov took a lead role in managing international relations for *OAO Biopreparat*. In December 1991, for example, Shcherbakov participated in a Soviet delegation which visited the US Army Medical Research Institute of Infectious Diseases (USAMRIID) at Fort Detrick, Maryland and other sites associated with historical US BW programmes. Shcherbakov at this time is reported to have known some English and to have a "*passing acquaintance with U.S. affairs*". Upon his return he is reported to have co-written a summary with Kalinin which claimed that the observations of the visiting delegation proved the continued existence of the US BW programme.[82] In March 1994, Shcherbakov also participated in a Russian inspection visit of Evans Medical Limited in Speke, Merseyside. It can be concluded, in the light of his overseas travel, that Shcherbakov is one of the most trusted members of the *OAO Biopreparat* hierarchy and that, in addition, he may have been tasked with making assessments of biological capabilities in Western countries.

In 2006, *OAO Biopreparat*'s Scientific-Technical Administration was involved in the process of vetting ISTC proposals to ensure that no state secrets would be released as a result of collaborative work. According to Smithson, some uncertainty existed at this time as to whether *OAO Biopreparat* required all, or only selected, bioscience institutes to submit their proposals for this pre-screening process, which took anywhere from two to four months.[83] If successful, the institute received an official letter from *OAO Biopreparat* confirming that the project did not relate to dual-use technologies or relate to information which was classified as a state secret. The ISTC office in Moscow subsequently terminated its activities on 15 July 2015 after Russia's withdrawal from the agreement.[84]

A Centre for Export Control was also established within *OAO Biopreparat*, headed by Valentin Evstigneev, the first deputy director of the joint stock company. In June 2002, the Centre became incorporated into a system of independent expert identification. This system has been developed to aid exporters with the identification of controlled items and technologies which are on the Russian export control lists. Although the exporter has responsibility for the identification of controlled items, there is also the right to transfer this obligation to an organization that has received special permission to undertake this activity under a procedure established by the Russian Federation government. The expert organization then bears the responsibility for the correct and substantiated results of the identification of controlled goods and technologies. According to a Russian government white paper on WMD non-proliferation, the presence of this system is "*an important legal enforcement tool*" and the "*expert support of customs and control agencies is aimed at expanding capabilities for keeping track of exports of 'sensitive' goods and technologies in order to prevent their unauthorized transfer*".[85]

The export control centre was granted permission to perform identification of a range of human, animal and plant pathogens, genetically modified microorganisms, toxins, equipment and technologies originally listed in a Presidential decree of 8 August 2001 devoted to biological export control. It should though be noted that the Centre does not perform this role with regard to technologies for the *production* of pathogens, toxins or genetically modified organisms which are subject to export control in the original decree.[86] *OAO Biopreparat* was also at this time able to procure an export licence within one week compared with one month for the Russian Ministry of Health.[87] The establishment of the Centre for Export Control allowed *OAO Biopreparat* to retain influence with regard to the operations of its former facilities at Obolensk and Kol'tsovo, where, for example, it was the licensing authority with regard to the export of microbial strains (see immediately below).

Estimating Numbers of *Biopreparat* Personnel

Since the collapse of the USSR, there has been an apparent substantive downsizing in the size of the headquarters staff maintained by *OAO Biopreparat*. Soviet archive documents reveal that in 1988, the Main Administration *Biopreparat* employed at least 91 such personnel within 11 separate departments (1st to 11th) and a so-called Second Administration. A more recent entry in an online directory stated that 36 headquarters staff were employed by the company.[88] By June 2006, *OAO Biopreparat* employed within its headquarters, a total of just 17 persons, including secretarial staff. Shcherbakov's Scientific-Technical Administration employed just two members of staff. There appears to have been some natural wastage of personnel (retirement, etc.), but the figures, if accurate, do appear to be indicative of a considerable downsizing in the scale of operations conducted by the company and possibly reflect its diminished status within the Russian biopharmaceutical sector.

Estimating the staffing numbers with regard to *Biopreparat*'s operations in its entirety is, given the paucity of reliable data, an extremely difficult task. Recent studies with regard to staffing levels at *Biopreparat* during the late Soviet period suggest that the organization employed anywhere from 30,000 to 40,000 personnel, of whom about 9000 were scientists and engineers.[89] In the period since the collapse of the USSR, estimating staffing levels is made more problematic by the complex structures which have evolved with regard to the control and management of facilities within the company's network. Vladimir Davidov, a deputy general director of "*Biopreparat*", claimed in September 1993 that his organization employed 100,000 personnel. This figure must be considered as highly unreliable.[90] Possibly the best source of information relating to numbers of personnel in the post-Soviet period are two brochures issued by *RAO Biopreparat*. The first, produced some time before April 2001, reports that the company employed around 6000 scientists and specialists working in the fields of microbiology, biotechnology, genetic engineering, immunology and biophysics. Of these, one was a Full Member and one was a Corresponding Member of the Russian Academy of Sciences, one was a Corresponding Member of the Russian Academy of Medical

Sciences, and 16 were Academicians of the branch academies. The company at this time employed 36 professors, 70 Doctors of Science and 519 Candidates of Science (equivalent to PhDs).[91] Meanwhile, a second *RAO Biopreparat* brochure produced some time between April 2001 and March 2004 stated that more than 36,000 personnel were engaged in production for the company.[92] A very rough estimate of the numbers employed within *Biopreparat*-controlled/affiliated facilities in the late 1990s to early 2000s is therefore approximately 42,000 personnel.

However, although there appears to have been initially no appreciable downsizing in total numbers of personnel employed by *Biopreparat*-affiliated facilities since the Soviet period, the company does appear to have lost significant numbers of scientists and engineers. By April 2005, Grigorii Shcherbakov was able to inform members of the NATO Sub-committee on the Proliferation of Military Technology (Science and Technology) that *OAO Biopreparat* employed approximately 5000 scientists. This represented a decrease of 1000 compared with the figures in the pre-April 2001 brochure.[93] At least part of the reduction in the scientific workforce may be accounted for by the emigration overseas of *Biopreparat* researchers. According to Andrew Weber, a key specialist at the US Defence Department, by 1999 around 300 scientists from the organization had emigrated to the US, Europe and elsewhere, "*but no one knows how many have moved to countries of BW proliferation concern*".[94] The recent transfer of the Obolensk and Kol'tsovo research centres to *Rospotrebnadzor*, and of the Institute of Highly Pure Biological Preparations, to the *FMBA* (see below), suggests an additional sharp reduction in the current numbers of scientific personnel employed in facilities affiliated to *Biopreparat*.

Scientific Meltdown: The Loss of Core Components of the *Biopreparat* Network

Shortly after the creation of the *OAO Biopreparat*, there emerged evidence that the new company had been forced by the Russian government to relinquish any remaining control of its core applied R&D facilities. The first indication that a major reorganization of two of *OAO Biopreparat*'s most significant research institutes was underway occurred on 31 March 2005 with the announcement that *Vektor* would in future be known as the Federal State Enterprise of Public Health "State Scientific Centre of

Virology & Biotechnology '*Vektor*'". More significantly, it was also announced that the facility was to be subordinated to the Federal Service for Supervision in the Sphere of Protection of Consumer Rights and Wellbeing of the Person (*Federal'naya sluzhba po nadzoru v sfere zashchity prav potrebitelei i blagopoluchiya cheloveka* or *Rospotrebnadzor*) headed by Genadii Grigor'evich Onishchenko.[95] This organization incorporates the closed and highly secretive network of anti-plague institutes and their associated satellite stations. Just six months later, on 26 September 2005, Russian Prime Minister Mikhail Fradkov signed Order No. 1514-r for the creation at Obolensk of the Federal Budgetary Science Institution "State Scientific Centre of Applied Microbiology and Biotechnology" (*Federal'noe byudzhetnoe uchrezhdenie nauki "Gosudarstvennyi nauchnyi tsentr prikladnoi mikrobiologii i biotekhnologii"* or *FBUN GNTs PMB*). The facility was also placed under the control of *Rospotrebnadzor*.[96]

The restructuring of the management of facilities at Obolensk and Kol'tsovo may have been part of a wider strategy being implemented by the Russian government to restrict the handling and storage of highly dangerous pathogens to facilities belonging to the closed networks managed by the Russian Ministry of Defence and *Rospotrebnadzor*. Onishchenko was now reported to exercise close control over the operations of facilities at Obolensk and Kol'tsovo and his approval was required for visits to Obolensk and for any international visits undertaken by his scientists. He also acted to appoint his own senior scientists from *Rospotrebnadzor*'s secretive anti-plague network as directors. It may be the case that as a result, much closer links were formulated with both the anti-plague facilities and the Russian Ministry of Defence.

One of the reasons for the transfer of the *Biopreparat* facilities in Kol'tsovo and Obolensk to *Rospotrebnadzor* may have been the reports of serious financial irregularities which emerged in the national media at this time. The Russian newspaper, *Rossiiskaya gazeta*, had reported on its findings in December 2001. It revealed that during the early 1990s, Major General (Reserves) Nikolai Nikolaevich Urakov (see Fig. 8.3), head of the *Biopreparat* applied microbiology facilities in Obolensk, had been actively participating in the writing off and subsequent sale of equipment. Among items disposed in the "*hundreds of tonnes*" are reported to have been drying facilities, container tanks and fermenters. According to the report, "*Entire production lines, including imported ones, disappeared without trace. The tracks of the money obtained can today be uncovered only by the law-enforcement organs*".[97] Urakov is also reported to have been

Fig. 8.3 Major General (Reserves) Urakov, *Traktir* Restaurant, Serpukhov, 13 February 2001 (©A. Rimmington)

responsible for the transfer of one of the Centre's buildings, Building No. 82 into private hands. *Rossiiskaya gazeta* noted that "*a unique pilot installation for the manufacture of genetically engineered insulin, whose creation cost enormous sums obtained for the most part from the state budget and which is located in building No. 82, also passed into private ownership*".[98]

Urakov is reported to have founded more than 20 small enterprises at the Obolensk site which are reported to have paid him and his associates a "land tax". These included a company which bottled vodka. One media outlet reported that in 1997, during a routine inspection by the Moscow *oblast'* tax police, officers discovered a tanker with Georgian number plates in a nearby car park. It was carrying a full load of alcohol. The company's management denied any connection with the vehicle and no documentation could be discovered relating to it.[99] A Western visitor to the site in 2001 noted that of 100 buildings originally belonging to *GNTs PMB*, only 2, Building No. 1 and Building No. 8, remained under its control. Some of the others were rented to private companies and most of the others were simply closed down. *Rossiiskaya gazeta* reported that Urakov appeared to have amassed his own personal fortune and "*has built for himself several luxury houses*".[100] Leitenberg and Zilinskas report that *Rospotrebnadzor* ordered an audit of *GNTs PMB*'s accounts which revealed "*irregularities*".[101] In 2003, Urakov was reported to have been removed as director at Obolensk, allegedly because of his past mismanagement of the facility.[102] A temporary administrator, Vladimir Plekhanov, was appointed to bring the centre back to solvency, with Vladimir Yakovlevich Volkov, the former first deputy director, initially taking over responsibility for scientific affairs at the centre.[103] In January 2006, Ivan Alekseevich Dyatlov, a former deputy director at the *Mikrob* Anti-Plague Institute in Saratov, was placed in charge of the facility.[104] Despite the allegations which arose against him, Urakov is reported to have still been a regular visitor to facilities at the Obolensk site. He died on 28 April 2017 following a long illness and was buried in Russia's Federal Military Memorial Cemetery.[105]

Another facility where *Biopreparat*'s control appears to have been greatly diminished is the Federal Budgetary Science Institution "State Scientific Centre of Virology and Biotechnology *Vektor*" (*Federal'noe byudzhetnoe uchrezhdenie nauki "Gosudarstvennyi nauchnyi tsentr virusologii i biotekhnologii"* or *FBUN GNTs VB*—see Fig. 8.4). At the end of September 2005, Onishchenko, the head of *Rospotrebnadzor* (see above) issued an order appointing Professor Il'ya Gennadievich Drozdov as director of *Vektor*. Drozdov had previously been a deputy director at *Rospotrebnadzor*'s closed *Mikrob* anti-plague institute (Saratov). Since July 2001, he had also served on the Russian Ministry of Industry, Science and Technology's Expert Council on Questions of Biosecurity. Sandakhchiev, who had been the director for almost 23 years, remained in place as an advisor to Drozdov until his death from heart disease on 26 June 2006 (see Fig. 8.5).[106] A key feature of Drozdov's new regime at *Vektor* appears to have been a clampdown on overseas contacts, with foreigners being prohibited from entering Building No. 1, scientists being forbidden to attend international meetings and a suspension of applications to the International Science and Technology Centre (ISTC).[107] A clear demonstration of this new clampdown was provided when two individuals, including *Vektor*'s deputy director for commercial affairs, were arrested in Siberia for disclosing state secrets to the US.[108]

Fig. 8.4 Road Sign for State Scientific Centre of Virology and Biotechnology *Vektor* (Kol'tsovo Novosibirsk *oblast'*), Photo 74815454 (© Evgeniy Muhortov | Dreamstime.com)

Leitenberg and Zilinskas allege that under Drozdov's new leadership, institutional corruption at the *Vektor* facility became blatant. They cite an open letter, originally quoted in the 7 April 2010 edition of the Russian newspaper *Novaya gazeta*, directed to President Vladimir Putin by Pavel Korchagin, a local official in Siberia. Korchagin specifically alleges Drozdov's participation in a criminal group which had succeeded in winning a contract for the rebuilding of *Vektor*'s HIV research laboratory. The group, he alleges, then embezzled more than 140 out of 183 million roubles, which had been allocated to them for the project. Korchagin pointed to the fact that in the past two years under Drozdov, 9 Doctors of Science and 20 Candidates of Science had departed the *Vektor* facility. He also argued that in such a situation, Russia could be left defenceless against dangerous viral diseases. The letter appears to have had the desired effect and shortly thereafter, on 18 July 2010, Drozdov and an administrator closely linked to him were dismissed from *Vektor* for financial misdealings and irregularities.[109]

Fig. 8.5 Monument to Academician Lev Stepanovich Sandakhchiev (Director, State Scientific Centre of Virology and Biotechnology *Vektor*, 1979–2006), Kol'tsovo, Photo 100540294 (© Evgeniy Muhortov|Dreamstime.com)

There is an extraordinary twist to this tale, detailing the misdemeanours of Drozdov. For, on 20 January 2017, under the sensational headline "*Biological weapon terror fears after Russian scientist in charge of producing Ebola and smallpox for the Kremlin disappears*", the UK newspaper, *The*

Daily Mail, reported that the scientist had vanished without trace. The Russian Interior Ministry confirmed that he was at this time on their wanted list and faced immediate detention if he could be found. He was also put on Interpol's wanted list by the Russian authorities.[110] Local news sources report that after his dismissal, Drozdov had travelled to Saratov and then disappeared. He has been accused by state investigators of misappropriating some 2 million roubles—at the time worth around US$55,000. The microbiologist allegedly used the cash to purchase an apartment. According to Elena Chernyayeva, a regional deputy prosecutor, the criminal case against Drozdov was filed against him in 2014, and a Novosibirsk court had given its approval for his arrest in absentia.[111] There appear to have been no sightings of Drozdov since.

In January 2011, *Rospotrebnadzor* appointed as successor to Drozdov, Professor Aleksandr Nikolaevich Sergeev (Doctor of Medical Sciences). Initially appointed as acting director at *Vektor*, on 10 March 2011 he was confirmed permanently in this position.[112] Born on 5 November 1953, he had graduated in 1977 from the Second Moscow State Medical Institute. He then joined *Biopreparat*'s *VNIIPM* in Obolensk. In 1980, he completed his graduate studies at the *Mikrob* anti-plague institute in Saratov. Later, in 1981, he was transferred to *Vektor* where a succession of promotions led to him being appointed as a head of department and then in October 2005 as a deputy general director for scientific and epidemiological work. For a short period from 2009 to 2010, he then joined the Russian Ministry of Health's *Mikrogen* Association, concerned with vaccine manufacture. Given his long-term involvement as a reputable and leading scientist at *Vektor*, Sergeev's appointment has presumably been very welcome news to the researchers and other staff members of the Kol'tsovo centre.

Another key R&D facility formerly subordinate to *OAO Biopreparat* has been subject to restructuring of its management. On 8 July 1997, the government of the Russian Federation issued decree No. 846 which granted the State Scientific-Research Institute of Highly Pure Biological Preparations (St Petersburg) the status of a state scientific centre.[113] On 27 April 2005, the institute, employing 270 scientific personnel, was transferred to the control of the Federal Medical-Biological Agency (*FMBA*).[114] On the same date, the *FMBA* also took over control of the State Scientific-Research Institute of Biological Instrument-Making (Moscow)—see above.[115]

Operating in a System Subject to Endemic Corruption and Organized Crime

The cases of high-level corruption at Obolensk and Kol'tsovo detailed above are of considerable cause for concern. For, with their sophisticated research and manufacturing capabilities, combined with their enormous holdings of highly dangerous pathogens, these facilities would make tempting targets for those seeking to acquire biological weapons. It is therefore perhaps worthwhile to briefly examine the environment in which *Biopreparat* had to operate in the period immediately after the dissolution of the USSR. A number of authors report the existence of systemic corruption in the 1990s which operated at the highest levels of government and industry alongside an organized crime sector which subsequently emerged as one of the strongest actors within the FSU. In 1996, some official Russian estimates calculated that "*organized crime may now control some 40,000 state and private organizations and over half of all economic entities within the FSU*".[116] A little over a decade later, in January 2007, Oleg Safonov, Deputy Minister of the Interior, stated that 446 organized criminal groups with a complex internal structure were at that time operating in Russia. Of these, nine had international links and were being tracked by the Ministry's department with responsibility for combating organized crime and terrorism.[117] Aleksandr Yelin, Deputy Head of the Interior Ministry's Department on Organized Crime, also warned in January 2007 that Russian mafia organizations had wrested control of hundreds of major industrial companies and were trying to gain political influence.[118] Meanwhile, in March 2007, Rashid Nurgaliev, head of Russia's Interior Ministry, revealed in a leaked speech that almost one-tenth of the Russian Federation was controlled by organized crime groups, often operating with little or no official resistance. A specialist in organized crime, Aleksei Mukhin, commented that Russia at this time could have been home to 10,000 criminal groups employing up to 300,000 people.[119]

There was also an explosion in criminal activity within Russian military and security organizations with an "*extensive and expanding overlap between Russian "power ministries" and the criminal world*".[120] A report by Turbiville Jr, covering the period from 1991 to 1996, argued that, "*institutionalized military crime is now flourishing in Russia. Military crime and corruption are associated directly with the Russian Ministry of Defence (MOD), the General Staff, and other senior staffs; military transportation, construction, and logistic organizations; combined arms units and commands; technically oriented and highly trained strategic strike and air defence formations; military research organizations; and military-educational components. Individual military criminals range in rank from general and field-grade officers to the newest conscripts, while organized criminal groupings within the military work independently and in partnership with 'commercial' ventures and outside civil sector 'mafias' at home and abroad*".[121] Turbiville concluded that, "*given the systemic criminality in other Russian branches of service, there is substantial reason to question whether military personnel responsible for Russian nuclear, chemical, and biological weapons are more fundamentally reliable than the demonstrably corrupt military officials assigned to responsible positions elsewhere*".[122]

Meanwhile, evidence of widespread corruption in the Russian pharmaceutical industry has been detailed in a study published by Vacroux in 2004. Initially, she notes, entrepreneurs focused on the more lucrative economic sectors of the Russian Federation, such as banking and exportable raw materials and succeeded in gaining control of ministries with responsibility for these sectors. Attention then shifted to remaining opportunities in less substantive markets such as the pharmaceutical industry. Vacroux argues, "*Placing their representatives near or at the top of important federal and regional bureaucracies allowed these entrepreneurs to capture some of the key regulators of their industry and to lobby the remaining parts of the government for benefits 'from within'*".[123] Such corrupt networks within government agencies have become the norm consisting of three elements: "*a commercial or financial structure, a group of bureaucrats to protect them, and a tie with law enforcement officials*".[124]

The criminality prevalent within the pharmaceutical industry at this time was highlighted in December 2005 when one of the heads of the Open Joint Stock Company "Irbit Chemical-Pharmaceutical Factory" (Irbit, Sverdlovsk *oblast'*) was arrested. According to the Federal Service for Narcotics Control, an illicit laboratory discovered at the facility had been engaged in the production of illegal drugs—about 5 kilogrammes of amphetamines per week—which were sold in Moscow.[125] A more recent indicator of the endemic corruption within the sector was the arrest on 17 November 2006 of Andrei Taranov, Director of the Russian Ministry of Health's Federal Fund for Mandatory Medical Insurance (FFMMI).[126] He, together with seven senior colleagues, was accused of accepting bribes from Russia's Territorial Mandatory Health Insurance Funds—regional counterparts of the FFMMI—and also from pharmaceutical manufacturers and suppliers in order to secure their participation in the state programme for supplemental drugs.[127] Speaking in Ekaterinburg a few days later, Dmitrii Medvedev, at this time Russia's First Deputy Prime Minister, asserted: "*The state of affairs in the pharmaceutical industry is disgusting....Order must be restored there with the use of various means, including use of criminal procedures*".[128]

Recombinant Insulin Production in Kazakhstan: *Biopreparat*'s Ambitious Plans to Convert Military Biological Facilities at Stepnogorsk

In April 1993, Stephen Asbridge, a senior sales executive at the now defunct UK company, LH Fermentation Ltd. (Stoke Poges), received a highly unusual enquiry regarding the supply of fermenters and associated equipment to a remote location in Kazakhstan. Asbridge had worked on biotechnology projects in the Soviet Union for more than a decade and knew all the major players in the sector, but the Kazakh Science Production Complex *Biomedpreparat (Kazakhstanskii nauchno-promyshlennyi kompleks Biomedpreparat* or *KNPK Biomedpreparat*) was totally unfamiliar to him. Intrigued by the potential scale of the proposed project, Asbridge hastily arranged a meeting in Moscow in June 1993 with Aleksandr Petrovich Mikhailichenko and Aleksandr Nikolaevich Kosinov, both at that time deputy directors at the Kazakh facility. The two scientists revealed to Asbridge that *Biomedpreparat* was located at a remote site on the Kazakh steppe at Stepnogorsk in Tselinograd *oblast'*. They reported that the complex had formerly been called the Stepnogorsk Scientific Experimental-Industrial Base (*Stepnogorskaya nauchnaya opytno-promyshlennaya baza* or *SNOPB*) and had participated in *Biopreparat*'s programme—see above. During the subsequent discussions, the two Stepnogorsk scientists indicated that the management of the *Biomedpreparat* complex had formulated a two-stage production programme based on anticipated funding from the Kazakh government in combination with investment from foreign organizations and biopharmaceutical companies. The first part of the programme was aimed at the construction of facilities for the production of antibiotics, steroids and enzymes. The second part focused on the production of recombinant human insulin.[129]

It was the latter project concerning the production of recombinant human insulin which was of most interest to *Biomedpreparat*'s ambitious director, Colonel (Reserves) Genadii Nikolaevich Lepeshkin. The Stepnogorsk scientists reported to Asbridge that the technology for the process had been originally developed by Sergei Mikhailovich Navashin's facility in Moscow, the All-Union Scientific-Research Institute of Antibiotics (*Vsesoyuznyi nauchno-issledovatel'skii institut antibiotikov*—or *VNIIantibiotikov*) and subsequently modified by Lepeshkin's own scientists. The Russian State Archive of the Economy (*RGAE*) reveals that the project to produce recombinant human insulin in Stepnogorsk dates back to July 1988, when the USSR Ministry of the Medical and Microbiological Industry (*Minmedbioprom*) issued an Order relating to the organization of industrial production of the hormone. This was followed on 17 February 1989 by a second Order, No. 82, issued by *Minmedbioprom* and signed by the minister, Bykov, which again focused on setting up the industrial production of genetically engineered human insulin. This new Order specifically spelt out the key role to be played by *Biopreparat* and its first deputy director, Alibekov, in the scaling up of the *VNIIantibiotikov* technology. It also referred directly to the participation of Lepeshkin in the project and the siting of production in Stepnogorsk with an anticipated capacity of 100 kg of substance per annum.[130]

Biomedpreparat revealed to Asbridge that they had succeeded in scaling up the recombinant human insulin process technology in 20 m^3 fermenters. Only "tens of grams" of insulin had been produced in Stepnogorsk thus far and the intention was to scale up production to 100 kg per annum as per the original *Minmedbioprom* Order. The estimated total demand in Kazakhstan at this time was 50 kg per annum with Russian demand for this product estimated at 400 kg. In order to further scale up the process, the Stepnogorsk scientists indicated that they wished to undertake the reconstruction of Building No. 221, which had been the main production facility at *SNOPB*. During a subsequent visit to the facility in Stepnogorsk, Asbridge was informed that the building had been constructed in 1982. It incorporated an outer clean zone and an inner operational zone which housed ten fermenters (20,000 litre capacity). Entry and exit to and from the production area took place via a special decontamination room. The manufacturing process was observed to descend from the upper to the lower floors and terminated in the underground floors. Service areas were located between the floors.[131] Asbridge's observations correspond closely to the account provided by Leitenberg and Zilinskas who indicate that this building comprised of six stories with two of them underground. They reported that it incorporated twenty 1000-litre fermenters on its top floor and ten 20,000-litre fermenters on the lower floors and was a crucial component in the mobilization plant's capability to produce around 300 metric tonnes of weaponized *B. anthracis* spores per ten-month production cycle.[132] Lepeshkin wished to purchase equipment for a complete production line in Building No. 221, including a fermentation plant and downstream processing facilities. The overall aim at this time was to develop a GMP facility which as well as possessing the capability to produce recombinant insulin, also had maximum flexibility with regard to its range of products.[133]

During the meeting in Moscow, the Stepnogorsk scientists reported that the fermentation component of the *Biomedpreparat* project had an estimated projected cost of around £2 million. The total area available for installation of the modular fermentation lines in Building No. 221 was 18 × 50 m^2.[134] They reported that as well as approaching LH Fermentation they had also sent enquires to New Brunswick (US) and an unidentified Japanese company. The original preferred configuration which they outlined for the projected facility was ten 10-litre, seven 100-litre and three 1000-litre fermenters. However, following the discussions with LH Fermentation, the *Biomedpreparat* personnel opted for a combination of ten 10-litre, seven 100-litre and three 1000-litre fermenters in conjunction with LH's high-pressure cell disruption system. The representatives from *Biomedpreparat* also indicated that they would require a four-year supply of spare parts for any equipment ordered since they could not guarantee future allocations of hard currency. Another major component of the insulin project concerned the purchase of equipment for downstream processing in clean room facilities and formulation of the final product into tablets, phials and creams. An unknown German company had been engaged in negotiations with representatives from *Biomedpreparat* concerning the supply of appropriate equipment. It was also planned that quality-control and environmental-control facilities would be constructed as part of the project.[135]

During the course of the Moscow meeting it was revealed by the Stepnogorsk scientists that *Biopreparat*'s Moscow-based State Institute for the Design of Enterprises of the Biological Industry (*Giprobioprom*) had been selected as the main contractor for the reconstruction of *Biomedpreparat*'s Building No. 221. This institute, which also had a branch based in Kol'tsovo (Novosibirsk *oblast'*), was, at the time, one of

Biopreparat's main design facilities. *Giprobioprom* was clearly at the very heart of the recombinant human insulin project in Kazakhstan and a key condition for the purchase of equipment for Stepnogorsk was that *Giprobioprom* personnel would receive extensive training from any supplying biotechnology companies in the West. Stephen Asbridge left the Moscow meeting in June 1993 convinced that the UK's newly introduced export controls to prevent the proliferation of biological weapons technology were highly likely to prohibit his company's participation in the project.[136]

Although LH Fermentation subsequently withdrew their interest, some support for the Stepnogorsk insulin project did come from the government of Kazakhstan which offered a pledge of funding during a visit by President Nursultan Nazarbayev. Although Kazakhstan had declared its independence on 16 December 1991, the *SNOPB* facilities in Stepnogorsk had in fact remained under *Biopreparat* administrative control well into 1992. During this "*transition phase*", all of the classified documentation held at the plant was repatriated to the Russian Federation along with some sensitive equipment such as an octagonal test chamber.[137] Nazarbayev and his colleagues were now eager to stamp their authority on the Stepnogorsk site and, as detailed above, reorganized *SNOPB* in 1993 as the Kazakh Science Industrial Complex *Biomedpreparat*.[138] Shortly thereafter, on 16 November 1993, the Kazakh Cabinet of Ministers issued Decree No. 1140 for the creation of the National Centre for Biotechnology of the Republic of Kazakhstan (*Natsional'nyi tsentr po biotekhnologii Respubliki Kazakhstan* or *NTsB* for short). *Biomedpreparat* was incorporated within *NTsB* which was based in Stepnogorsk.[139] Initially, Professor Murat Kurmashevich Gilmanov was appointed head of *NTsB* but he was shortly thereafter replaced by Lepeshkin. In his new position, the latter was tasked with heading an ambitious national civil biotechnology programme in Kazakhstan incorporating more than 200 individual scientific projects being conducted at 25 R&D facilities.[140]

The lack of funding rapidly emerged as the key issue at Stepnogorsk. The payment of salaries and other budgetary items from Moscow was suspended following the break-up of the Soviet Union and it was not until the end of 1992 that Kazakhstan's government provided an emergency payment of 50 million roubles to the facility. This was followed by an additional 90 million roubles funding for the first quarter of 1993.[141] However, the situation again deteriorated and in July 1995 it was reported that even salaries to senior scientific personnel had been left unpaid for several months.[142] This dire financial backdrop, with *Biomedpreparat* also left unable to pay utility and other bills, had a major impact on staffing levels at the facility. By August 1994, many staff had returned to the Russian Federation and the number of personnel had been reduced by 50 per cent to give a total workforce at this time of just 450 people.[143] Of those remaining, only 60 or so were qualified scientific staff and engineers.[144]

Nazarbayev did eventually publish a special decree of Kazakhstan's Cabinet of Ministers on "*The Manufacturing of Medical Preparations in Stepnogorsk*" which earmarked—but eventually was unable to deliver—US$25 million investment for *Biomedpreparat*, the bulk of the money being intended for the insulin project. It was in the wake of this announcement that some very preliminary work was undertaken on the creation of a genetic engineering block within Building No. 221. The eventual intention was to establish a small pilot production unit to scale up production of recombinant human insulin. It occupied a small area of approximately 50 m^2. By the time an official UK trade and industry visit was undertaken to the facility in July 1993, construction of one clean room within the block had recently been completed. The walls were tiled, although the finish was observed to be of relatively low quality. Four recombinant pilot-scale vessels with capacities of 100 to 150 litres were observed at this time in storage under plastic wrapping.[145] However, despite the decree and Lepeshkin's own personal commitment to the scheme and the establishment of a line of credit in a European country, *Biomedpreparat* were subsequently unable to secure adequate investment from Kazakhstan's government to pursue the insulin project.[146]

The lack of funding from Kazakhstan's government also had a detrimental impact on *Biomedpreparat*'s implementation of its range of other civil programmes. In 1991, for example, the facility had begun industrial production of an immobilized protease called *Profezim*.[147] This preparation was originally developed by researchers at the USSR Academy of Sciences Siberian Division's Institute of Cytology and Genetics (Novosibirsk) and *Biopreparat*'s Scientific-Research Design Institute of Biologically Active Substances (Berdsk).[148] *Profezim* accelerates the wound-healing process and had been found to be especially effective in the treatment of burns, festering wounds, trophic and radiation ulcers and chronic osteomyelitis. One of the key customers for the preparation in the recent past had been the USSR Ministry of Defence who used it in the treatment of battlefield casualties.[149] Demand for *Profezim* and analogous preparations significantly exceeded supply and plans were drawn up for the establishment of an independent production unit at *Biomedpreparat.* However, a key limiting factor for the expansion of production had been the lack of specialized sterilization equipment and consequently immobilized enzymes had to be sent to Moscow for treatment. In order to address the problem of sterilizing these preparations and other pharmaceuticals and medical products (such as disposable syringes), an important new project was initiated at the complex. The construction of a radioisotope sterilization building, which used cobalt-60 for irradiation with g-rays, was completed in 1994. It was intended that the new facility would be operated in conjunction with the *Medpolimer* enterprise in Pavlodar.[150]

Another project initiated at *Biomedpreparat* had been the reequipment of existing facilities for the manufacture of disposable syringes. It was intended that the first production line would start in 1994 with an annual capacity of 25 million syringes. Subsequently, it was intended that production would be scaled up to 150 million syringes per annum.[151] However, delays in financing again put this project behind schedule and production was not ultimately initiated. Again, in September 1994, it was reported that local scientists had prepared plans to open a new facility at *Biomedpreparat* for the production of drugs from plants. Sufficient capacity was reported to exist but funds would be required to reequip and convert part of the enterprise. Apparently, Lepeshkin, Marat Sovetovich Koishibaev, the head of administration of Stepnogorsk (*Stepnogorskaya goradministratsiya*), and Shimanaev, President of the Joint Stock Company *Progress*, were all receptive to the idea, but once again the project floundered because of a lack of adequate financing. Later, in 1996, representatives of *Biomedpreparat* contacted a UK-based distributor with a request that they source second-hand tablet presses with a total value of US$100,000. However, when the equipment was located and offered for sale to the Stepnogorsk representatives, they were unable to proceed with the purchase because the promised line of funding did not materialize.[152] It is clear that while the Stepnogorsk scientists were eager and willing to pursue civil pharmaceutical programmes, these were destined to remain pure pipedreams when faced with the economic realities of newly independent Kazakhstan.

US Engagement with *Biomedpreparat* and Destruction of BW Facilities at Stepnogorsk

It was only with the subsequent involvement of the US that reliable funding streams finally became available to the scientists based in Stepnogorsk. In December 1993, the US and Kazakhstan signed an agreement to convert *Biomedpreparat* and redirect personnel to peaceful activity, such as producing pharmaceuticals. One of the first collaborative projects to be launched by the US in conjunction with Stepnogorsk scientists focused on antibiotics production at the site. Under the umbrella of the Co-operative Threat Reduction Programme, *Biomedpreparat* was selected as a participant in an industrial partnerships scheme. The US Defence Special Weapons Agency (DSWA), which was responsible for procurement of appropriate Western companies for participation, selected Allen & Associates International Ltd. (AAI), based in Washington, as *Biomedpreparat*'s partner. AAI was run by John Allen, reportedly a former intelligence agent who had operated in Laos and Cambodia and who had good political connections.[153] DSWA awarded a US$5.82 million cost-sharing contract with the US contributing $2,685,992, and AAI an additional US$3,139,620.[154]

Details of the US investment in *Biomedpreparat* were announced by Harold Smith (Special Assistant to the US Defence Secretary, William Perry) at a news conference in April 1995 in Almaty held jointly with Sagdat Nurmagambetov, Kazakhstan's Defence Minister. The collaborative project was centred on a joint venture, subsequently registered in March 1996, named Kamed Resources. It aimed to utilize 1000–2000 m^2 of a building previously associated with biological weapons production, for the manufacture, packaging and distribution of vitamins, pharmaceuticals and antibiotics. In May 1996, the project was given a major boost when the international company ICN Pharmaceuticals, Inc. (US) linked up with AAI in the conversion of the Stepnogorsk plant. In addition to financial support, ICN were to contribute equipment, materials expertise and training.[155] However, despite this crucial injection of industrial expertise, the first project aimed at the production of antibiotics collapsed in 1997 amid recriminations on both sides.[156] Whatever the rights and wrongs of this episode, it does clearly illustrate the severe difficulties involved in creating, at such a remote location, commercially viable bioscience companies able to produce pharmaceuticals to acceptable international standards.

In 1996, Kazakhstan agreed to the dismantlement of *Biomedpreparat* in cooperation with the US Department of Defence. An original "Weapons of Mass Destruction Infrastructure Elimination" (WMDIE) agreement had previously been signed on 3 October 1995 by the US Department of Defence and Kazakhstan's Ministry of Science—Academy of Sciences which focused on the provision of an initial US$6 million for the elimination of nuclear test tunnels at Semipalatinsk. An amended agreement was subsequently signed between these two organizations on 7 June 1996 providing additional funding to a total of US$22.5 million and permitting usage of WMDIE funds for closure of nuclear test holes at Semipalatinsk and the dismantlement of the infrastructure for BW production at Stepnogorsk.[157]

The successful execution of the dismantlement project was later highlighted at a joint conference held in Stepnogorsk in July 2000. Brian Hayes, a representative from the Defence Threat Reduction Agency (DTRA), described the project as an "*extreme success*". According to him, "*The third out of four phases (elimination of all BW infrastructure such as electricity, water and sewage pipes etc.) had been completed three weeks before the conference. Phase four would now be started, that is complete elimination of buildings 221, 231 and 600, designated for production, desiccation and R&D, respectively*".[158] However, the implementation of the project was not without problems, including the fact that there was a lack of sophisticated lifting equipment which meant workers could only proceed with the use of simple tools.[159] Zilinskas reports that he witnessed some of this work in 2003 and notes that, eventually, buildings 211, 221, 231 and 600 were completely demolished.[160]

The dismantlement of the infrastructure at *Biomedpreparat* represents another major success—alongside that achieved at *Sibbiofarm* in Berdsk (see below)—for the US in its efforts to prevent the international proliferation of BW technologies and expertise. It had succeeded in eliminating the most advanced and sophisticated BW production complex the world has ever seen. The government of Kazakhstan meanwhile, should also be congratulated for its early recognition that it could not safely retain and operate the facilities at Stepnogorsk. Given its own lack of finance and relevant expertise, any attempt at wholesale conversion of the remote site was a long way beyond its own capabilities in this regard. Clearly, such was the scale of the weapons complex that even advanced Western nations would have found such a project a formidable and enormously expensive challenge.

Biopreparat's Glittering Legacy: The Emergence of International Biotechnology Reagent Companies

It is beyond the scope of the present work to document the fate of each and every research and production unit which was controlled by *Biopreparat*. However, it is worth pointing to the spectacular success which was achieved by the organization with regard to restriction enzymes. It is notable that the commercialization of *Biopreparat*'s capabilities in this area was achieved by scientists in Lithuania where the desire to break out of the Soviet strangle-hold restricting scientific innovation was at its strongest. Nevertheless, this example makes a useful case study and highlights the fact that newly privatized companies in the post-Soviet states could draw on large-scale state investments from the Soviet era to potentially compete internationally and create multi-million-dollar businesses.

During the 1970s, *Biopreparat* was faced with the problem of providing an adequate supply of restriction enzymes, which are vital for genetic engineering work, for researchers working within its ambitious civil and military programmes. The first tentative step towards initiating production of these reagents was the assembly of a national collection of strains producing restriction enzymes by the Moscow-based All-Union Scientific-Research Institute of Genetics and Selection of Industrial Microorganisms (*VNIIgenetika*) in collaboration with the N.F. Gamaleya Institute of Epidemiology and Microbiology (Moscow).

Meanwhile, the Soviet authorities were about to take a decision which would eventually have a massive impact on the development of the USSR's capabilities in the area of isolation and production of restriction enzymes. For, on 1 March 1975, with the aim of creating a facility specializing in the development and production of both highly purified restriction enzymes for molecular biology research and industrial (bulk) enzymes, *Biopreparat* established the All-Union Scientific-Research Institute of Applied Enzymology (*Vsesoyuznyi nauchno-issledovatel'skii institut prikladnoi enzimologii* or *VNIIPE*) in Vilnius, Lithuania. The institute was created on the basis of the pre-existing Central Research Laboratory of the Vilnius Experimental Enzyme Plant. Some 100 staff were transferred from the lab to the newly created *VNIIPE*. Additional personnel were recruited from local biochemical, chemical, botanical, chemical technological institutes and Vilnius University's Department of Chemistry and so on. By 1 January 1976, the *Biopreparat* institute had a staff of 194 people, including 136 research personnel. The Soviet archives reveal that the number of personnel at *VNIIPE* then increased from 299 in 1982 to 333 in 1985.[161] This figure again grew and by 1989, some 739 people were employed at the facility.[162] The first director appointed in 1975 was Dr Antanas Glemža. In January 1978, he was replaced by Dr Donatas Kazlauskas, who in turn relinquished his position in 1987 to Dr Henrikas Dudėnas.[163]

VNIIPE initially comprised 8 departments and 21 laboratories. The departments were: Department of Genetics and Selection of Microorganisms (three laboratories); Department of Physiology and Biochemistry of Microorganisms (three labs); Department of Mathematical Modelling (one lab); Department of Enzyme Purification Technologies (four labs); Department of Physical-Chemical Methods (four labs); Department of Environmental Protection (two labs); Department of Enzyme Technologies and their Application (three labs); and Department of Scientific-Technical Information (one lab).[164] The institute was also the lead facility in the Science Production Association *Ferment* (*NPO Ferment*), which incorporated both the institute itself and a number of linked manufacturing plants.

The Lithuanian institute received a major boost with the appointment of Major General (Reserves) Anatolii Andreevich Vorob'ev as first deputy director of *Biopreparat*, the post later occupied by Alibekov. Vorob'ev had been a major figure within the network of BW facilities managed by the USSR Ministry of Defence. His last posting in the military system had been as deputy director at the Scientific-Research Sanitary Institute (*67-i km* settlement, Zagorsk *raion*).[165] Vorob'ev is reported to have been married to a minor *Biopreparat* functionary, Lyudmila I. Petrova who served as *VNIIPE*'s "curator" (see above). According to Zilinskas and Leitenberg, this meant she supervised the Lithuanian facility, but had no technical responsibility for its work. She was, apparently, the only employee in Vilnius who was trusted with knowledge of *Biopreparat*'s role in military programmes.[166] Presumably owing to his wife's position at *VNIIPE*, Vorob'ev became a frequent visitor to Lithuania and is reported to have channelled more than US$10 million in hard currency to the institute for the purchase of sophisticated Western equipment.[167] Vorob'ev appears to have developed close links to Henrikas Dudėnas, the General Director of *NPO Ferment* from 1987 to 1991. Vorob'ev's memoirs record that he personally awarded the Transferable Red Banner award to Dudėnas in Vilnius in 1986.[168]

The origins of *VNIIPE*'s interest in restriction enzymes lay with one of its outstanding scientists, Professor Arvydas Janulaitis. He is a "*stocky man who looks more like a Chicago Bears linebacker than a biochemist*"[169] and appears to have been *Biopreparat*'s first and most successful biotech entrepreneur. He was born on 26 September 1942 in the Lithuanian industrial town of Panevezys and then graduated from the Department of Biochemistry at Leningrad State University in 1969. He subsequently completed his Candidate's and Doctoral degrees in biological sciences. He joined *VNIIPE* in 1975 and served initially as head of the Genetic Engineering Laboratory.[170] It was in this laboratory that Janulaitis first isolated strains producing well-known restriction enzymes and technologies for their purification. The enzymes were supplied both to *Biopreparat*, for its applied work, and to research facilities engaged in basic research. In recognition of the crucial importance of his laboratory to the development of Soviet molecular biology, it was in 1978 reorganized as the Department of Molecular Biology Enzymes. Janulaitis was appointed head of the new department and later was promoted to the position of deputy director of the institute.[171]

In his new position at the institute, Janulaitis also embarked on collaboration with *VNIIgenetika* (see above) and by 1980 had collected data on 62 strains producing restrictases, determining their optimal rates of growth, favourable storage conditions and so on. As early as 1976, he was ready to sell his first restriction enzymes to Soviet laboratories. However, Janulaitis reports that at that time, to complete sales to just one institute required an astonishing amount of paperwork. Therefore, he took the decision to simply give his enzymes away and "*within a few months, everyone knew my name*".[172] It was the demonstration of *VNIIPE*'s capabilities in this area that subsequently allowed Janulaitis to put pressure on the Soviet authorities for an allocation of hard currency to purchase fermentation equipment from Germany for its new pilot plant (see below). By 1983, Janulaitis was also seeking permission from the Soviet authorities to contact companies in the West. It was his intention to sell restriction enzymes in bulk to the companies which could then be resold under their own brand names. Astonishingly, Janulaitis succeeded in gaining permission and eventually secured orders, via the Soviet foreign trade organization Licenzintorg, from a Japanese company. None of the precious hard currency which *VNIIPE* earned was returned to the institute but Janulaitis viewed the interaction with the West as a learning process, including the gaining of an understanding of international commercial negotiations, which would stand him in good stead in the future.[173]

One of the key problems facing researchers at *VNIIPE* was the lack of a pilot plant for production of purified enzymes developed at laboratory scale. It utilized the Vilnius Experimental Enzyme Plant for developing technologies with regard to its bulk enzymes R&D programme but had access to no equivalent pilot production for its purified enzymes. Therefore, in 1984, it began the construction of its own pilot plant equipped with fermenters up to 1000 litres in size.[174] This facility came on stream in 1987. The first large-scale production of restriction enzymes for research establishments in the Soviet Union was subsequently organized at the pilot plant. However, in the first instance the project turned out to be something of a white elephant and "*it turned out that the purified enzymes intended for the (sic) enzymatic analysis, which were planned to be produced, were not in such a demand as had been projected*".[175] After long discussions, it was eventually decided to redirect the pilot plant to the manufacture of recombinant proteins for use in the Soviet health service. The plant subsequently began the production of alpha2b-interferon and in 1990 the drug preparation was registered in the USSR as Reaferon™. The project was considered to be politically important at this time as it could be used to give credence to the argument that the Soviet Union was employing cutting-edge biotechnologies in the pharmaceutical sector.[176]

By 1987, Janulaitis was able to register the pilot production facility as a semi-independent company. A year later the CPSU politburo granted the company permission to sell his restriction enzymes directly to Western clients and to retain the profits from this interaction. In 1989, Janulaitis was appointed director of *VNIIPE* and immediately took the decision to refocus the bulk of its resources on restriction enzymes. According to Steven Dickman, a reporter from *Science*, "*this would require lots of ingenuity and hard work, but little costly equipment. And if they could be successfully made, they could be sold in small quantities and easily shipped—just the product for a small state with aspirations for independence from Moscow*".[177]

Janulaitis' institute began to request that its researchers, when on holiday in remote locations across the Soviet Union, collect soil samples containing extremophilic bacteria, which might possess unique enzymes. In addition, he used his extensive network of friends and colleagues based at research institutions and hospitals across the USSR to send him samples from their localities.[178] Prior to the collapse of the USSR, the facility had screened more than 25,000 microbial strains and was producing around 83 restriction enzymes and 26 DNA modification enzymes. *VNIIPE* rapidly became one of the world leaders in this area of scientific research and by 2000 it is estimated that it had accounted for around 30 per cent of the more than 3000 restriction enzymes which had been isolated across the globe. One of the outstanding achievements claimed by the institute concerns the first discovery anywhere in the world in 1983 of the minor DNA base (N^4-methylcytosine, m^4C).[179]

On 11 March 1990, Lithuania became the first Soviet state to announce a restitution of independence. The Soviet authorities responded with the imposition on 20 April of an economic blockade with the cessation of the supply of raw materials, especially oil, to Lithuania. The face-off between the Lithuanian and Soviet authorities placed "All-Union" establishments such as *VNIIPE*, which had been designed to serve demand for the whole of the Soviet Union in a most precarious position. It remained to be seen whether the institute could be funded by Lithuania alone for its yearly budget amounted to more than 50 per cent of the country's annual expenditure on biotechnology R&D. Alibekov reports that he was at first ordered to suspend funding of the institute and then this decision was reversed. One crucial factor in the Soviet leadership's calculations regarding *VNIIPE* was the fact that it was at the forefront of Soviet research into the production of genetically engineered products. Using a 1000-litre fermenter, its pilot plant had recently produced 1 million doses of recombinant alpha-2 interferon which were despatched for packaging to pharmaceutical factories in the city of Kaunas (see above). As Alibekov suggests, if the plant had shut down, then "*our top Party officials would lose some of the high-quality medical care they expected*".[180]

In December 1990, the author was a visitor to *VNIIPE* in Vilnius. The journey to the Lithuanian capital had been a long and difficult one. It was impossible at this time to book flights to Vilnius, so recourse had to be made to the Tallink ferry which linked Helsinki to Tallinn in Estonia. This proved to be a very rough winter crossing of the Baltic Sea. Upon eventual arrival in Vilnius, the political tensions and the impact of the economic sanctions were everywhere apparent. Just five weeks later, on Sunday 13 January 1991, Soviet tanks, led by members of Alpha Group, an elite KGB special forces unit, attacked pro-independence demonstrators at the television tower, killing 14 people.[181]

It was during this acutely difficult period that the management of the institute—now renamed the Institute of Applied Enzymology Fermentas—took the decision to continue to deliver supplies to customers in the Soviet Union. They recognized that Russia and Ukraine were likely to remain the key market for their enzymes and that any cessation in supply would do irreparable damage to the reputation of the institute. This far-sighted decision ensured that in the decades to come, Fermentas remained a major player in the reagents market of the Former Soviet Union. By 1991, *VNIIPE*, with a staff of 700 and an annual budget of 12.3 million roubles was reported to account for 70–80 per cent of restriction enzymes used in the Soviet Union, although cooperatives were becoming increasingly competitive in the fight for a slice of the Soviet market.[182]

Fermentas had, in the period from its creation through to the independence of Lithuania, developed a list of enzymes which is described as being larger and more diverse than any other in the world at this time. The institute began selling these products in bulk to six or seven companies.[183] By September 1992, orders were being received from 15 companies based in Europe, the US and Japan, including US-based New England Biolabs. The annual income in 1991 from these sales was reported to be US$340,000.[184]

In January 1992, the facility was recognized as a state research institute by the Lithuanian government and it was renamed the Institute of Biotechnology Fermentas.[185] In 1995, the AB Fermentas company, specializing in the production of more than 300 products including restriction endonucleases (over 140 items), DNA and RNA modification enzymes, kits for investigations into molecular biology, nucleic acids and their components and so on, was spun-off from the Institute of Biotechnology. By 2000, the company, under the leadership of Professor Viktoras Butkus, employed 130 people including 40 research staff. It was at this time exporting products to the US, Japan, Germany, the UK, France and others and had trade representatives in some 45 countries.[186] In 2003, a controlling enterprise, MBI Fermentas Inc., was established in Burlington, Canada, and subsequently developed a global distribution network for its molecular biology products. The latter included reagents for nucleic-acid and protein purification, restriction and modifying enzymes, molecular weight markers, and other life science research and diagnostic tools. By 2009, the company is reported to have generated full-year revenues of approximately US$54 million. Fermentas International was acquired by Thermo Fisher Scientific in 2010 for US$260 million in cash.[187]

It was in the face of the ongoing political turmoil within the Baltic republics that the Soviet authorities became concerned that the supply of critically important restriction enzymes from Lithuania might be interrupted and place *Biopreparat*'s military and civil R&D programmes in jeopardy. It was these considerations which presumably led to the decision to create a second major centre in the USSR for the production of restriction endonucleases for the Soviet biotechnology industry. This was established at the Scientific-Research Design Institute of Biologically Active Substances (*Nauchno-issledovatel'skii konstruktorsko-tekhnologicheskii institut biologicheski aktivnykh veshchestv* or *NIKTI BAV*) in Berdsk, outside Novosibirsk, Russia. By 1989, *NIKTI BAV*'s parent organization, the Science Production Association *Vektor* (itself also part of *Biopreparat*), was able to supply 32 restriction enzymes to Soviet researchers. Some of *Vektor*'s restriction enzymes were also exported internationally via Oxford Virology plc (see above) and tests in the UK revealed that their quality was excellent.[188]

The Russian restriction enzymes business appears subsequently to have been transferred to the Science Production Association (*NPO*) *SibEnzim*—later renamed *SibEnzaim*—which was created in 1991 by Professor Sergei Kharitonovich Degtyarev.[189] The *NPO*—which was subsequently incorporated within SibEnzyme Ltd.—was established on the basis of two laboratories, the Laboratory of New Enzymes and the Laboratory of New Bacterial Producers, both belonging to *Biopreparat*'s Science Production Association *Vektor*. The latter retained a 9.4 per cent stake in the new company with 14 unnamed Russians holding the remaining 90.6 per cent. The Berdsk biologics factory was utilized for the production of the reagents. The company then appears to have been transferred to the Russian Academy of Sciences Siberian Division's Novosibirsk Institute of Bioorganic Chemistry, since renamed as the Institute of Chemical Biology and Fundamental Medicine. As of 2013, SibEnzyme was reported to have a staff of 45 of whom 25 were engaged in production. There is also a Scientific-Research Group which is reported to be engaged in the following activity: searching for producers of new enzymes (restriction endonucleases and other enzymes of nucleic exchange); characterization of new enzymes; development of methods of application of new enzymes in genetic engineering; and the cloning of enzyme genes and the study of their properties.

SibEnzyme remains the most significant Russian producer of reagents for molecular biological research. It continues to produce enzymes and other preparations for molecular biology and biotechnology. As of 2013, the organization was distributing a catalogue containing some 170 restriction endonucleases, some of which are unique internationally. Most of the enzymes found in the company's catalogue have been discovered and characterized by NPO SibEnzyme. The organization at this time possessed three distribution centres in Moscow, St Petersburg and Novosibirsk and was marketing its preparations internationally under the SE trademark via New England Biolabs. Its enzymes were also being marketed by Sigma, Roche Diagnostics, Promega and so on.

Russia's Civil Focus for Facilities at Vostochnyi

Following the collapse of the USSR, two new joint stock companies were established on the basis of the Omutninsk Chemical Factory. The principal company was the Joint Stock Company *Vostok* (AO *Vostok*) which was initially under *RAO Biopreparat*. The choice of the name *Vostok* is related to the fact that during the Soviet period, the Omutninsk Chemical Factory had been the head organization of the Production Association *Vostok* and the name was therefore familiar in the local biotechnology industry. In addition, the Joint Stock Company "Omutninsk Chemical Factory" (see Fig. 8.6) was created as part of *Vostok* and Sergei Leonidovich Zhukov was initially appointed general director.[190]

A key figure in the management of the Vostochnyi complex at this time was Vladimir Arkad'evich Valov. He had served as director during the Soviet period (see above) and continued to occupy this position through to at least the mid-1990s. According to personnel from Vostochnyi, during the post-Soviet chaos and dislocation, the autocratic style and local political clout of Valov were instrumental in preventing any degradation of the plant's capability and its technological base. Indeed, by the mid-1990s the facility was reckoned to be one of the strongest in Russia's biopharmaceutical industry. Perhaps most significantly in terms of the future activity at the plant is the fact that even as late as 1995, there had been no appreciable brain-drain of personnel and equipment and production lines at the plant had been successfully mothballed.[191]

During the early 1990's Valov took the decision to re-focus production on the manufacture of disposable syringes using equipment reportedly purchased by *Biopreparat* from Werner Kammann Maschinenfabrik GmbH (Germany). The Omutninsk factory emerged at this time as Russia's largest producer of high-quality disposable syringes with output in 1995 expected to reach 330 million packs.[192] Not surprisingly manufacture of this product emerged as the company's core business. While engaging only 10 per cent of the total workforce in their manufacture, sales of the syringes accounted for 90 per cent of the facility's profits. The Russian

Fig. 8.6 Exhibition Stand of Joint Stock Company Omutninsk Chemical Factory, Russian Healthcare Exhibition, Moscow, December 1997 (©A. Rimmington)

Federation's State Statistical Committee noted that the company's output had increased by 44 per cent in the first ten months of 1995 compared to the same period the previous year.[193]

It was Valov's intention to use the profits from the new core business—the manufacture of disposable syringes—to relaunch microbiological production at the plant. He had already been successful in obtaining government support for the facility's incorporation within the State Amino Acids Programme and intended using technology developed by the All-Russian Institute of Genetics and Selection of Industrial Microorganisms (Moscow) to begin the production of these preparations.[194] In the medium term, Valov intended to import semi-finished pharmaceuticals for formulation and packaging in clean rooms built for the Soviet Union by the South Korean company Samsung.[195] The pharmaceuticals were then intended to be distributed to his sales branches based at Barnaul, Ekaterinburg, Krasnodar, Moscow, Omsk, Omutninsk, Perm, Tomsk, Ufa, Vladimir, Volgograd, Minsk (Belarus) and Almaty (Kazakhstan) which possessed their own storage warehouses.[196] His long-term aim was for the Omutninsk facility to produce the semi-finished pharmaceuticals itself and then for the branches to formulate and package these preparations.[197] However, Valov's old-fashioned leadership style appears to have not sat comfortably with the new owners after privatization and around this time he left for Kirov where he headed a regional industrial administration.

In the period after Valov's departure, the production of disposable syringes continued to grow, with output in 2003 standing at 480 million syringes per annum. At the time this represented one-third of all syringes being manufactured within the Russian Federation. The continuing commercial success of the plant meant that of the 2500 personnel formerly employed at the Omutninsk Chemical Factory, 1700 were still working at the plant in July 2003. Equipment at this time at the Omutninsk Chemical Factory had also been maintained in good working order with fermentation capacity at the plant comprising twelve 63 m^3 fermenters, with the remaining fermenters having a capacity of 15 m^3. Similarly, ultra- and microfiltration equipment at the factory was at this time being maintained in good condition. The plant also had access to limited animal testing facilities, primarily rabbits, concerned with safety testing of its syringes.

The other major facility at the Vostochnyi site during the Soviet period was the Omutninsk Scientific Experimental-Industrial Base (*Omutninskaya nauchnaya opytno-promyshlennaya baza* or *ONOPB*)—see above. At some point in the early 1990s it became an open joint stock company. The general director of the facility at this time appears to have been Yurii Leonidovich Mironenko. *ONOPB* had previously been focused on scaling up technologies for subsequent transfer to the contiguous Omutninsk Chemical Factory. However, after the collapse of the Soviet Union, it was increasingly forced to undertake production of products itself and since 1992 had been engaged in the production of antitumour drug substances. In 1995 for example, it succeeded in producing the first batch of bleomycin, an antitumour antibiotic used in the treatment of cancer. The preparation is subsequently reported to have been exported to Germany. According to Mironenko, speaking in January 2001, *ONOPB* had no significant debts and its financial position compared favourably with other industrial enterprises in the Kirov region.[198]

US Assistance in Dismantling of *Biopreparat* Dual-Use BW Facilities in Russia

The single-most important biological conversion project implemented with US assistance in the Russian Federation concerns that pursued at the private limited company, *Sibbiofarm* Production Association (*OOO Proizvodstvennoe Ob"edinenie Sibbiofarm* or *OOO PO Sibbiofarm*). *Sibbiofarm* incorporates the Berdsk Factory of Biological Preparations (formerly known as the Berdsk Chemical Factory) in Novosibirsk *oblast'*, which as detailed above was one of the Soviet Union's main reserve mobilization BW facilities. On 22 June 2006, Frank Record of the US Department of State described the Berdsk biologics plant as, at that time, constituting "*the largest dual-use biological production facility in Russia*".[199] *Sibbiofarm* also incorporated the Small State Agricultural Enterprise "Druzhba" (Nizhnyaya Kamenka, Novosibirsk *oblast'*).[200] *Sibbiofarm* was included in the list of entities which were transferred on 4 April 1991 to the ownership of the State Concern *Biopreparat*.[201] *Sibbiofarm* was privatized in 2004 and is now a private limited company.

After the collapse of the USSR, the production facilities at Berdsk are reported to have "*slid into dereliction*" and left its staff, many of whom were scientists, laid-off and facing the prospect of destitution. Valerie DeThomas, a senior programme manager at CRDF Global, states that during this period prior to its privatization "*it was basically a dying enterprise...People were laid off, and a lot of those people were scientists with advanced degrees*".[202] The economic and social crisis afflicting the plant was put into sharp focus when, on 28 April 2004, there was a serious incident involving Aleksandr Salostii, Director of *Sibbiofarm*. He was attacked with a blunt object as he left his apartment to go to work. According to the local publication, *Novaya sibir'*, the incident was being viewed as an act of hooliganism although the law-enforcement agencies had not ruled out that the attack was linked to Salostii's activities as head of *Sibbiofarm*.[203]

Sibbiofarm was subsequently given a new lease of life when it was privatized in 2004. The new owners and investors had ambitious plans to access global markets for consumable biologicals, such as a fertilizer and antibiotics. However, the new consortium lacked both the technical skills and knowledge required to operate in their new capacity as a private business. They therefore enlisted the help of CRDF Global in training *Sibbiofarm*'s staff, which was reported to include nine former weapons scientists, renovating its facilities, and developing a sound and sustainable business plan.

According to DeThomas, "*We worked with them to help develop plans for a new, revived, and modern production facility.... We also helped identify the areas where they lacked expertise or understanding of how markets work outside of Russia, and then enlisted subject matter experts to help develop business plans for individual units*".[204] CRDF Global's interaction with *Sibbiofarm* was supported with significant funding from the US Department of State's BioIndustry Initiative (BII).[205]

The *Sibbiofarm* project is reported to have consisted of two phases. In the first, CRDF Global assisted with the procurement of all kinds of equipment and renovation of the plant.[206] In the second phase, CRDF Global assisted with the provision of additional training of *Sibbiofarm* staff with regard to the use of the equipment procured in the first phase. Equipment manufacturers travelled to Berdsk to deliver training and *Sibbiofarm* staff also travelled to the manufacturers premises to receive instruction.[207] In addition, for three weeks in July 2012, five *Sibbiofarm* scientists, accompanied by two translators, spent three weeks based at Michigan State University (MSU) and the nearby Michigan Biotechnology Institute (MBI). The principal goal of the MSU and MBI programme was to train the Russian scientists to use new equipment. There was also a strong focus on learning the practical aspects of designing, operating and maintaining a modern fermentation pilot plant and, crucially, on approaches which can be taken to bridge the gap between early innovations and commercial applications.[208] The Soviet Union had wrestled unsuccessfully with this latter problem for many decades. According to Jim Wolfram, an independent consultant, who assisted CRDF, "*This training simulated, as closely as possible, the production conditions that exist at the Russian company....A permit was secured to allow their non-pathogenic organism to be transported to the United States. Therefore, the conditions of the fermentations and assays were familiar to the trainees....During the training, a new process control variable was tested. The result yielded a 50 per cent gain in production of the product per unit volume. A positive increase like this, if reproducible at the Russian plant, could be financially significant in their sustainability*".[209]

Besides facilitating the large-scale upgrades to the plant's infrastructure, a key part of the near-decade interaction between CRDF Global and *Sibbiofarm* concerned the removal and destruction of its dual-use BW facilities.[210] These were perceived to pose "*inherent security and health risks for the city of Berdsk, the Novosibirsk region, and internationally for both Russia and the U.S.*"[211] The BW elimination project is reported to have been completed sometime around 2007 or thereafter, in accordance with the objectives of the US State Department's BioIndustry Initiative. Several large fermenters capable of producing "*material of defence interest*" were dismantled.[212] Some buildings may also have been demolished, but thus far no details of the operation have been released either by Russia or the US.

Sibbiofarm was later judged by CRDF Global to be both profitable and sustainable and the facility has now emerged as one of Russia's largest producers of commercial biologics, manufacturing more than 30 different products including fertilizers and antibiotics for livestock. There are also plans by the company, which is now ISO 9001: 2015 compliant, to access overseas markets including Europe, North America and South America. According to Aleksandr Krichevskii, General Director of Sibbiofarm, "*Our experience with CRDF Global and the BioIndustry Initiative has been successful and productive. The programme provided a creative and benevolent atmosphere for our employees that allowed them to get answers to all their questions*".[213]

The importance of the interaction between US non-proliferation agencies and *Sibbiofarm* cannot be overstated. The outcome can be considered to be a major collaborative success for both parties. The Russian company has emerged from the process as a modern biologics company able to compete at both national and international levels. Its highly skilled workforce has been retained and enhanced through a rigorous training programme. From a national, state-level point of view, the success of the project means that decades of significant Soviet investment have not been simply thrown away and the revitalized facility can now be considered a major growth pole for the Novosibirsk region. For the US, the project has demonstrated both to the Russian government and to the local scientific community that it is a reliable partner and will pursue agreed collaborative projects through to their successful conclusion. In addition, it has succeeded in eliminating significant dual-use biological facilities which had been in existence for more than half a century.

Ending Not with a Bang but a Whimper

The *Biopreparat* open joint stock company, which remained in existence from 2000 to 2010, was clearly a pale shadow of its former self which had initially bestrode the newly emerging Russian biopharmaceutical sector like a colossus. In the decade since the collapse of the Soviet Union, the profitable parts of its empire had been privatized and were now firmly

Fig. 8.7 Biotechnopark Platform for Innovative Pharmaceutical Companies (Kol'tsovo, Novosibirsk *oblast'*), Photo 100540263 (© Evgeniy Muhortov | Dreamstime.com)

focused on commercial objectives. In addition, as detailed above, there had been elimination of weapons facilities at *SNOPB* (Kazakhstan) and *Sibbiofarm* (Berdsk). The Institute of Engineering Immunology (Lyubuchany) and State Scientific-Research Institute of Highly Pure Biological Preparations (St Petersburg) are both reported to be engaged in solely civil research under conditions of transparency. In addition, the State Scientific Centre of Virology and Biotechnology *Vektor* (Kol'tsovo) is reported to have been "*demilitarized*" and has emerged as one of Russia's leading centres for disease surveillance and research (Fig. 8.7).[214] There can be little doubt therefore that the power and influence of *Biopreparat* have been greatly diminished.

It remains open to question as to whether *Biopreparat*, which continues to be tasked with special functions, will have a significant role in the future with regard to biosecurity and related issues in Russia. There were, during the period from 2005 to2010, encouraging indications that the company was engaging with a number of Western institutions. For example, on 25

April 2005, Grigorii Shcherbakov represented *OAO Biopreparat* during discussions at the State Duma with Members of the NATO Parliamentary Assembly's Sub-Committee on Proliferation of Military Technology [Science and Technology].[215] Again, from 2007 to 2010, the UK's Foreign Office also assisted *Biopreparat* in maintaining a dialogue with the UK Health Protection Agency's (HPA's) Centre for Emergency Preparedness and Response (CEPR). The two sides agreed to form a joint working group for cooperation on biosecurity, biosafety and development of innovative strategic vaccines, innovation treatment modes and therapies.[216] In May 2006, *Biopreparat* also participated in a Russian-Finnish Joint Biotechnology Laboratory workshop, held in Turku, Finland.

It is beyond the scope of an academic work, such as this one, to take this epic story of *Biopreparat* any further. The company now resides in the shadows and no longer takes centre stage. It remains impervious to additional academic investigation. If it continues to have a role in Russian programmes, it is almost certainly only a supporting one. Its previous, most important role in the Soviet Union, of managing a network of R&D and production facilities, has all but been taken away. Most of the key personnel who participated in Soviet programmes are in any case deceased or have taken up more lucrative positions in the commercial pharmaceutical sector. Military biological programmes are now once again the sole preserve of the Russian Ministry of Defence.[217] We thus end not with a bang but a whimper. Whether *Biopreparat* continues to exist or not is a moot point. It may simply find itself to be increasingly irrelevant and to have no useful role to play in Russia's new, post-Soviet, landscape.

Notes

1. Brown, A., Kaser, M., Smith, G.S., *The Cambridge Encyclopedia of Russia and the Former Soviet Union*, p. 143.
2. Umnov, V., Opasnost' bakteriologicheskoi voiny ostaetsya, *Komsomol'skaya pravda*, 19 September 1992, pp. 11–12; Ukaz Prezidenta Rossiiskoi Federatsii o strukture federal'nykh organov ispolnitel'noi vlasti, *Rossiskie vesti*, No. 6(430), 11 January 1994, p. 1.
3. Nechaev was a retired Colonel-General of the Medical Services. During his period of service in Afghanistan, he worked as the chief surgeon in the Central Military Hospital in Kabul. During the period 1985–1987, he was chief surgeon of the Soviet forces based in Germany. From 1987 to 1988, he was deputy head of the S.M. Kirov Military Medical Academy.

He served as chief surgeon of the USSR Ministry of Defence from 1988 to 1989. Nechaev's military career is outlined in Yubilei, *Voenno-meditsinskii zhurnal*, No. 12, 1994, pp. 68–69; Ryan, M., Personalia and the current health service, *BMJ*, Vol. 306, 3 April 1993, pp. 909–911, https://www.bmj.com/content/bmj/306/6882/909.full.pdf, Accessed on 4 September 2020.

4. Nechaev, E.A., Scientific concept of provision of medicines and drugs to Russian population, *Vestnik Rossiiskoi Akademii Meditsinskikh Nauk*, No. 7, July 1994, pp. 5–10, translated in *JPRS Report, Science & Technology, Central Eurasia*, JPRS-UST-94-033, 9 December 1994, pp. 46–51.
5. Rossiiskie farmakologi svoimi razrabotkami mogli by zavoevat' mir…, Otchet s tret'ei (LXVI) sessii Obshchego sobraniya RAMN, *Meditsinskaya gazeta*, No. 15(5448), 25 February 1994, p. 1.
6. *Rossiiskoe aktsionernoe obshchestvo Biopreparat*, date and place of publication unknown.
7. *Rossiiskoe aktsionernoe obshchestvo Biopreparat*, Brochure produced by RAO "Biopreparat", date and place of publication unknown; RAO "Biopreparat" ob'yavlyaet predvaritel'nye itogi 2000 goda, http://medi.ru/news/01222bioprep.htm, Accessed on 21 July 2006.
8. O naznachenii predstavitelei gosudarstva v sovet direktorov RAO "Biopreparat", *Pravitel'stvo rossiiskoi federatsii rasporyazhenie*, No. 1315-r, 12 August 1994, http://law.optima.ru.
9. Litovkin, D., Interv'yu: Valentin Evstigneev: "Shtamm Eboly v Rossiyu privezli razvedchiki", *Yadernyi Kontrol'*, No. 4, July–August 1999, pp. 15–25.
10. Postanovlenie pravitel'stva rossiiskoi federatsii, No. 505, *Rossiskaya gazeta*, 8 May 1999.
11. These comments were made at a conference organized by *Rosmedprom* in conjunction with the Russian Medical and Pharmaceutical Associations. The deputy general director of *RAO Biopreparat*, M.I. Grigor'ev, who also attended the meeting, refused to comment on the situation. Kushnareva, M., Medpromyshlennost'—ne Zolushka, a lyubimaya doch, v Minzdravmedprome RF? Zametki s kollegii 18 avgusta s.g., *Meditsinskii biznes*, 7–8(13–14), 1995, p. 2.
12. Kondratev, V., Kak odin gospodin gosudarstvo perekhitril: V otechestvennoi meditsinskoi promyshlennosti poiskhodyat nynche strannye veshchi, *Komsomol'skaya pravda*, 7 December 2000, p. 14.
13. Butrin, D., Kompanii. Minzdrav vzyalsya za ozdorovlenie "Biopreparata", *Kommersant-Daily*, No. 061, 6 April 2001, p. 5, http://news.mosinfo.ru/news/01/KDE/04/data/k040671.htm; Starobin, P., The trials of a drug czar, *Business Week*, No. 3717, 29 January 2001, p. 20.
14. Butrin, D., *Kommersant-Daily*, No. 061, 6 April 2001, p. 5.

15. The earliest identified reference to *Rosmedprom* and to Kalinin's position as President is in June 1995. Farmatsevticheskomu proizvodstvu—gosudarstvennuyu podderzhku!, *Farmatsevticheskii vestnik*, No. 12(29), 15–30 June 1995, p. 1.
16. Ibid.
17. *Farmatsevticheskii vestnik*, No. 4(155), 1 February 2000.
18. Ibid.
19. In February 1998, the Interdepartmental Health Protection Committee accused foreign companies of threatening to undermine the country's medical industry by taking control of Russian pharmaceutical enterprises. *Interfax News Agency*, Moscow, in English, 1502 gmt, 27 February 1998, carried in *Summary of World Broadcasts*, Part 1, Former USSR, SU/3/65, 3 March 1998, p. B4; Russian Federation Ukase sets composition of Security Council Interdepartmental Health Protection Committee, *Rossiiskaya gazeta*, 21 May 1996, translated in *FBIS Report, Science & Technology, Central Eurasia, Life Sciences*, FBIS-UST-96-025, 21 May 1996; Statue on Interdepartmental Commission on Genetic Engineering, *Rossiiskaya gazeta*, 17 May 1997, translated in *FBIS Report, Science & Technology, Central Eurasia, Life Sciences*, 17 May 1997.
20. Leitenberg, M., The possibilities and limitations of biological weapons conversion: Personnel and facilities, p. 125 in Geissler, E., Gazsó, L., Buder, E., *Conversion of Former BTW Facilities*, NATO Science Series, Kluwer Academic Publishers, 1998.
21. Rimmington, A., From Offence to Defence? Russia's Reform of its Biological Weapons Complex and the Implications for Western Security, *The Journal of Slavic Military Studies*, Vol. 16, No. 1, March 2003.
22. Interview with former *Biopreparat* scientist, 3 December 2001.
23. Interview with former *Biopreparat* scientist, UK, April 2006.
24. Smirnitskii, N., Zaraznyi biznes: Generala ot gennoi inzhenerii, *Rossiiskaya gazeta*, 4 December 2001.
25. Pravitel'stvo utverdilo plan privatizatsii federal'nogo imushchestva, http://www.medmoskva.ru/news/Out.aspx?Item=34080, Accessed on 4 April 2006.
26. O provedenii Mezhgosudarstvennoi nauchno-prakticheskoi konferentsii i Koordinatsionnogo soveta po problemam sanitarnoi okhrany territorii gosudarstv – uchastnikov SNG, *Rospotrebnadzor* Order No. 123, 12 May 2006, http://www.rospotrebnadzor.ru/docs/order/?id=247, Accessed on 5 July 2006.
27. *Rossiiskaya gazeta*, First Edition, Moscow, 28 February 1992, translated in *Summary of World Broadcasts*, Part 1, Former USSR, SU/1317, 29 February 1992, p. C3/1.

28. *Reuters*, 22 June 1994, carried in *Chemical Weapons Convention Bulletin*, No. 25, September 1994, p. 22. Syutkin was formerly a Deputy Chairman of the Committee overseeing the Department concerned with compliance to the 1972 Biological Weapons Convention. Kuntsevich was dismissed as Chairman in April 1994. See Frolov, D., Lateral move: Generals come and go, chemical weapons remain, *Segodnya*, 9 April 1994, p. 2, translated in *JPRS*, JPRS-UST-94-010, 3 May 1994, p. 1; Yeltsin closes plague plants, *The Sunday Times*, No. 8851, 10 April 1994, p. 19.
29. The 1994 appropriation for the Committee was reported to be 735.9 million roubles. See *Rossiiskaya gazeta*, 6 July 1994, translated in *BBC Summary of World Broadcasts*, 9 July 1994, carried in *Chemical Weapons Convention Bulletin*, No. 25, September 1994, p. 24; *Russian CW Destruction Program President's Committee*, 22 July 1993, unpublished research document kindly provided by academic colleague.
30. Litovkin, V., El'tsin zapretil raboty po bakteriologicheskomu oruzhiyu. Eto znachit: on velis', a nas obmanyvali, *Izvestiya*, Moscow Edition, No. 99(23673), 25 April 1992, p. 1.
31. *Biological Warfare Compound #19*, PaineWebber Incorporated, HFT Corporation, No date or place of publication, p. 5. This document will have been generated at some point between the creation of the committee in February 1992 and Kuntsevich's dismissal as its Chairman in April 1994.
32. Interview with *III* scientist, February 2001.
33. *Biological Weapons: Effort to Reduce Former Soviet Threat Offers Benefits, Poses New Risks*, United States General Accounting Office Report to Congressional Requesters, GAO/NSAID-00-138, Washington, April 2000, p. 15.
34. Smithson, A., *Toxic Archipelago: Preventing Proliferation from the Former Soviet Chemical and Biological Weapons Complexes*, Report No. 32, The Henry L. Stimson Center, Washington, D.C., December 1999, p. 51.
35. Kondrat'ev, V., Kak odin gospodin gosudarstvo perekhitril: V otechestvennyi meditsinskoi promyshlennosti proiskhodyat nynche strannye veshchi, *Komsomol'skaya pravda*, 7 December 2000, p. 14.
36. A similar report appeared in Tyler, P.E., Russian Lab Storing Germs Faces Cutoff of Electricity, *The New York Times*, 7 April 2002, Section 1, p. 11, https://www.nytimes.com/2002/04/07/world/russian-lab-storing-germs-faces-cutoff-of-electricity.html, Accessed on 3 September 2020; and Tyler, P.E., An unpaid bill threatens germ factory in Russia: Electric company could end bacteria studies, *International Herald Tribune*, 8 April 2002.

37. Coker, M., Russian anthrax storage in peril, *Deseret News*, 19 October 2002, https://www.deseret.com/2002/10/19/19683978/russian-anthrax-storage-in-peril, Accessed on 3 September 2020.
38. http://www.regions.ru/article/any/id/1731569.html, 28 January 2005, Accessed on 28 January 2005; http://www.regions.ru/article/any/id/1811561.html, 17 May 2005, Accessed on 17 May 2005.
39. Government of Moscow, http://www.mosreg.ru/news/4/2005/6/10/12433.html, 10 June 2005, Accessed on 10 June 2005.
40. Kondrat'ev, V., Kak odin gospodin gosudarstvo perekhitril: V otechestvennyi meditsinskoi promyshlennosti proiskhodyat nynche strannye veshchi, *Komsomol'skaya pravda*, 7 December 2000, p. 14.
41. Ibid.
42. Stone, R., Russia, NIH float big plan for former Soviet bioweapons lab, *Science*, Vol. 291, 23 March 2001, pp. 2288–2289.
43. Http://www.orgitecture.com/crdf/events/events_show.htm?doc_id=292713, CRDF, Accessed on 12 September 2020.
44. Warrick, J., Lev Sandakhchiev, *The Washington Post*, 30 June 2006, https://www.washingtonpost.com/archive/local/2006/06/30/lev-sandakhchiev/d7fa8536-9c18-4966-8bde-be890d337a47/, Accessed on 12 October 2020.
45. Leitenberg, M., The possibilities and limitations of biological weapons conversion, p. 125.
46. *Controlling Dangerous Pathogens: A Blueprint for U.S.-Russian Cooperation, A Report to the Cooperative Threat Reduction Program of the U.S. Department of Defence*, U.S.-Russian Collaborative Program for Research and Monitoring of Pathogens of Global Importance Committee, National Academy of Sciences, Institute of Medicine, National Research Council, National Academy Press, Washington, D.C., 1997, p. 24.
47. Miller, Engelberg, Broad, *Germs*, pp. 220–221.
48. Rimmington, A., Russia: Microbiological Conversion, *Oxford Analytica East Europe Daily Brief*, 5 June 1997.
49. Roffey, R., Unge, W., Clevström, Westerdahl, K.S., *Support to Threat Reduction of the Russian Biological Weapons Legacy – Conversion, Biodefence and the Role of Biopreparat*, Swedish Defence Research Agency, FOI-R-0841-SE, April 2003.
50. Severin, E.S., Porodenko, N.V., Research Centre for Molecular Diagnostics and Therapy, *Biomedical Science*, Vol. 2, No. 5, 1991, p. 545.
51. Official website of the Izvarino pharmaceutical plant, http://www.izvarino-pharma.ru, Accessed on 9 June 2001.
52. Ivchenko, L., The Building 'D' affair. A private firm has privatized a major medical factory for a song, Trud, 28 January 1993, p. 2, translated

in *JPRS Report, Science & Technology*, Central Eurasia: Life Sciences, 16 April 1993, pp. 50–51.

53. http://hghltd.yandex.com/yandbtm?url=http%3A//vladimir.polevanov.ru/archives/000018.html&text=%E0%EF%F0%EE%EC%E5%E4&reqtext=%E0%EF%F0%EE%EC%E5%E4%3A%3A1819103916&dsn=580&d=675483, Accessed on 11 July 2006.
54. Ivchenko, L., The Building 'D' affair, *JPRS Report, Science & Technology*, Central Eurasia: Life Sciences, 16 April 1993, pp. 50–51.
55. *Biological Science and Biotechnology in Russia: Controlling Diseases and Enhancing Security*, The National Academies Press, Washington, D.C., 2006.
56. Zaderzhany 3 zhitelya Omska po delu o pokushenii na glavu NPO "Elevar", 26 July 2005, http://www.news-24.ru/hronic/a-9246.html, accessed on 6 April 2006; Raskryto ubiistvo stolichnogo predprinimatelya, *Kommersant'*, 27 July 2005, http://www.news-24.ru/hronic/a-9246.html, accessed on 6 April 2006.
57. *About Scientific Activities of the JSC "Institute of Immunological Engineering"*, Lyubuchany, 8 February 2001.
58. Leitenberg, Zilinskas, with Kuhn, *The Soviet Biological Weapons Programme*, p. 264.
59. Rimmington, *Who's Who*, p. 332.
60. Ibid.
61. Observations made during a visit to the Institute of Immunology (Lyubuchany) by Dr Anthony Rimmington, University of Birmingham, 16 May 1994.
62. Pchelintsev, S.Yu., *About Scientific Activities of the JSC "Institute of Immunological Engineering"*; Leitenberg, Zilinskas, with Kuhn, *The Soviet Biological Weapons Programme*, p. 266.
63. Fedorov, L.A., *Sovetskoe biologicheskoe oruzhie: Istoriya, Ekologiya, Politika*, International Social-Ecological Union, Moscow, 2005, http://www.seu.ru/cci/lib/books/bioweapon/, Accessed on 10 January 2007.
64. Information provided by *III* scientist, February 2001.
65. Leitenberg, Zilinskas, with Kuhn, *The Soviet Biological Weapons Programme*, pp. 702–703.
66. Interview with former *Biopreparat* scientist, February 2007.
67. Alcoa and Alcoa Foundation Sponsor First-Ever Comprehensive U.S. Exhibition of Russian Art, Businesswire, 2 June 2005, https://www.businesswire.com/news/home/20050602005381/en/Alcoa-Alcoa-Foundation-Sponsor-First-Ever-Comprehensive-U.S., Accessed on 9 September 2020.
68. Interview with former *Biopreparat* scientist, February 2007.
69. *Institut Immunologii, Lyubuchany, Moskovskaya oblast', SSSR*, Chekhov, date of publication unknown, p. 2.

70. Information supplied by scientist at the Institute of Immunology, Chekov *raion*, Moscow *oblast'*, May 1994.
71. Ibid.
72. *About Scientific Activities of the JSC "Institute of Immunological Engineering"*.
73. Interview with former *Biopreparat* scientist, February 2007.
74. Ibid.
75. Alibek, K., Handelman, S., Is Russia Still Preparing For Biological Warfare? *The Wall Street Journal*, 16 February 2000.
76. Information from Russian scientist, February 2001.
77. Rimmington, *The Biotechnology & Pharmaceutical Industry in Russia 2010*, pp. 42–43.
78. Perechen' strategicheskikh predpriyatii i strategicheskikh aktsionernykh obshchestv, Presidential Decree No. 1009, 4 August 2004, http://www.president.kremlin.ru/text/docs/2004/08/75174.shtml, Accessed on 9 October 2006.
79. Interview with former *Biopreparat* scientist, February 2007.
80. Interview with former *Biopreparat* scientist, April 2006.
81. Farmobozrenie dlya vsekh, No. 13, 18 February 2001, http://mednews.com.ua/issues/med/issue_013.htm, Accessed on 25 May 2006.
82. Alibek with Handelman, *Biohazard*, p. 242.
83. Smithson, A.E., *Toxic Archipelago*.
84. Withdrawal of the Russian Federation from the Agreement Establishing the International Science and Technology Centre, Official website of the ISTC, http://www.istc.int/en/article/119, 7 July 2015, Accessed on 16 September 2020.
85. *The Russian Federation and Nonproliferation of Weapons of Mass Destruction and Delivery Systems: Threats, Assessments, Problems and Solutions*, CNS translation of a Russian government white paper on non-proliferation policy, http://cns.miis.edu/pubs/other/rusfed.htm, Accessed on 28 July 2006.
86. The Centre for Export Control has not been given responsibility for technologies listed in sections 1.6.1–1.6.2 of the Decree. GTK RF: O Perechne organizatsii po provedeniyu nezavisimoi identifikatsionnoi ekspertizy (GTK, N 07-56/41346), http://customs.net.ru/index.phtml?t=3&id=6141, Accessed on 28 July 2006.
87 Presentation by *Biopreparat* scientist, December 2006.
88. See Biopreparat, http://www.chembz.ru/mcitem3_904.html, Accessed on 28 July 2006.
89. Leitenberg and Zilinskas state that "at least" 30,000 people worked within the *Biopreparat* system. The figure of 40,000 is estimated by

Smithson. Leitenberg, Zilinskas, with Kuhn, *The Soviet Biological Weapons Programme*, p. 164; Smithson, A.E., *Toxic Archipelago*.

90. Information provided by Vladimir N. Davidov of the State Concern "*Biopreparat*" to DTI OSTEMS Mission, Moscow, 22 September 1993. The findings of the OSTEMS team are presented in *OSTEMS Mission to Russia*, September 20–30, 1993 Mission Report, BioIndustry Association, London, 1994.
91. *Russian Joint Stock Company Biopreparat*, Brochure (1) produced by the company, Date and place of publication unknown.
92. *Russian Joint Stock Company Biopreparat*, Brochure (2) produced by the company, Date and place of publication unknown.
93. 25–27 April 2005—Visit to Moscow, Russian Federation by the Sub-Committee on the Proliferation of Military Technology [Science and Technology], http://66.102.9.104/search?q=cache:K9t4xmWJg3YJ:nato-pa.int/Default.asp%3FSHORTCUT%3D716+nato+parliamentary+biopreparat&hl=en&ct=clnk&cd=6, Accessed on 20 July 2006.
94. Tucker, J., Bioweapons from Russia: Stemming the Flow, *Issues in Science & Technology*, Spring 1999, http://www.issues.org/15.3/p_tucker.htm, Accessed on 20 July 2006.
95. Pod grifom "soglasovano", Tsentr Parlamentskikh Kommunikatsii "ParlKomm", http://www.parlcom.ru/index.php?p=MC83&id=1778, Accessed on 13 September 2006.
96. http://www.cnews.ru/newsline/index.shtml?2005/09/29/188110, Accessed on 29 September 2005.
97. Smirnitskii, N., Zaraznyi biznes: Generala ot gennoi inzhenerii, *Rossiiskaya gazeta*, 4 December 2001.
98. Ibid.
99. Mk-Novosti, 16 April 1997, http://demo.rco.ru/topsearch/hl.asp?URL=c%3A%5CResearchDemoData%5CTopData%5C04%5F1997%5C116061%2Ehtm&CONTEXT=+&INFOSYSTEM=topdemo, Accessed on 21 September 2006.
100. Smirnitskii, N., Zaraznyi biznes.
101. Leitenberg, Zilinskas, with Kuhn, *The Soviet Biological Weapons Programme*, p. 203.
102. Orent, W., *Plague: The Mysterious Past and Terrifying Future of the World's Most Dangerous Disease*, Free Press, New York, 2004, p. 226.
103. Leitenberg, Zilinskas, with Kuhn, *The Soviet Biological Weapons Programme*, p. 203.
104. Dyatlov Ivan A., http://pircenter.org/en/experts/396-dyatlov-ivan-a, Accessed on 4 September 2020.

105. Ushla epokha: umer sozdatel' poslednego nauchnogo gorodka SSSR N.N. Urakov, https://smitanka.ru/news/ushla_epokha_umer_sozdatel_poslednego_nauchnogo_gorodka_sssr_n_n_urakov_/, 28 April 2017, Accessed on 23 October 2020.
106. *NGS Novosti*, http://news.ngs.ru/more/17576.shtml, 7 October 2005, Accessed on 7 October 2005; *Russian Information Agency Regnum*, http://www.regnum.ru/news/665165.html, 26 June 2006, Accessed on 26 June 2006.
107. Leitenberg, Zilinskas, with Kuhn, *The Soviet Biological Weapons Programme*, p. 245.
108. Russia charges rocket scientists with spying for U.S., http://www.mosnews.com/news/2006/12/22/scientistspy.shtml, Accessed on 6 February 2007.
109. Leitenberg, Zilinskas, with Kuhn, *The Soviet Biological Weapons Programme*, pp. 245–246, 791 (note 104).
110. Stewart, W., Biological weapon terror fears after Russian scientist in charge of producing Ebola and smallpox for the Kremlin disappears, *MailOnline*, https://www.dailymail.co.uk/news/article-4140268/Russian-scientist-Ilya-Drozdov-disappears.html, 20 January 2017, Accessed on 9 September 2020.
111. Gertcyk, O., Ex-head of virology state research centre Vector vanishes: Interpol alerted, *The Siberian Times*, https://siberiantimes.com/other/others/news/n0852-ex-head-of-virology-state-research-centre-vector-vanishes-interpol-alerted/, 20 January 2017, Accessed on 9 September 2020.
112. Direktorom GNTs VB "Vektor" naznachen Aleksandr Sergeev, Naukograd Kol'tsovo, https://kolcovo.ru/content/news/index.php?news=2855, 18 March 2011, Accessed on 9 September 2020.
113. O prisvoenii statusa gosudarstvennogo nauchnogo tsentra Rossiiskoi Federatsii Gosudarstvennomu nauchno-issledovatel'skomu institutu osobo chistykh biopreparatov, Postanovlenie Pravitel'stva Rossiiskoi Federatsii, No. 846, 8 July 1997, *Rossiiskaya gazeta*, 6 August 1997, p. 5.
114. Istoriya, Official website of the State Scientific-Research Institute of Ultra Pure Biological Preparations, https://hpb.spb.ru/ru/institute/history, Accessed on 8 September 2020.
115. Perechen' nauchno-issledovatel'skikh organizatsii, podvedomstvennykh Federal'nomu mediko-biologicheskomu agentstvu, http://www.fmbaros.ru/struct/strpor.php?spp3=2, Accessed on 5 March 2007.
116. Turbiville, G.H., Executive Summary, *Weapons Proliferation and Organized Crime: The Russian Military and Security Force Dimension*, INSS Occasional Paper 10, Proliferation Series, USA Institute for National Security Studies, US Air Force Academy, Colorado, June 1996,

https://apps.dtic.mil/dtic/tr/fulltext/u2/a365528.pdf, Accessed on 14 September 2020.

117. 446 Organized Criminal Groups Operate in Russia – Police, *ITAR-TASS News Agency*, http://www.itar-tass.com/eng/level2.html?NewsID=11204689&PageNum=0, Accessed on 31 January 2007.
118. Russian Mafia Taking Over Big Business: Official, *AFP*, 31 January 2007.
119. Osborn, A., Interior Ministry: 'One-tenth of Russia Under Criminal Control', *The Independent*, 15 March 2007.
120. Turbiville Jr, G.H., Executive Summary.
121. Turbiville Jr, G.H., Weapons Proliferation and Organized Crime: Russian Military Dimensions, *Airpower Journal*—Special Edition, 1996, https://www.hsdl.org/?view&did=439487, p. 19.
122. Ibid., p. 20.
123. Vacroux, A., Regulation and Corruption in Transition: The Case of the Russian Pharmaceutical Markets, pp. 133–151 in Kornai, J., Rose-Ackerman, S., *Building a Trustworthy State in Post-Socialist Transition*, Palgrave Macmillan, New York, 2004.
124. Ibid.
125. Head of Russian Pharmaceutical Company Arrested in Illegal Drugs Scandal, http://www.mosnews.com/news/2005/12/20/pharmscandal.shtml, Accessed on 15 March 2007.
126. Putin Asks Health Minister to Assist Probe into Fraud, http://en.rian.ru/russia/20061127/56048733.html, 27 November 2006.
127. Corruption Scandal Unfolds with Arrest of Russian Mandatory Health-Insurance Fund Chief, Global Insight, http://www.globalinsight.com/SDA/SDADetail7563.htm, Accessed on 15 March 2007.
128. *ITAR-TASS news agency*, 22 November 2006.
129. The meeting with the *Biomedpreparat* scientists took place in Moscow on 4 June 1993.
130. Ob organizatsii proektirovaniya promyshlennogo proizvodstva genno-inzhenernogo insulina cheloveka, Order No. 82 of the USSR Ministry of the Medical and Microbiological Industry, Moscow, 17 February 1989, *RGAE*, Fond 788, Opis 1, Ed. Kh. 1167.
131. Observations made by Mr Stephen M. Asbridge during a visit to the *Biomedpreparat* Complex, Stepnogorsk, 14 July 1995.
132. Leitenberg, Zilinskas, with Kuhn, *The Soviet Biological Weapons Programme*, pp. 277–279.
133. Meeting with Aleksandr Petrovich Mikhailichenko (deputy director of *KNPK Biomedpreparat*) and Aleksandr Nikolaevich Kosinov (deputy director and head of laboratory at *KNPK Biomedpreparat*), Moscow, 4 June 1993.
134. Ibid.

135. Ibid.
136. On 10 March 1993, Asbridge had participated in a Foreign and Commonwealth Office seminar in the Locarno Suite on the introduction of legislation to control the export of biological materials and equipment that could be used for the proliferation of biological weapons.
137. Leitenberg, Zilinskas, with Kuhn, *The Soviet Biological Weapons Programme*, p. 280.
138. Korotko ob avtorakh (1994), *Stepnogorsk. Zdes' propisany nashi serdtsa.*, p. 202.
139. O merakh po obespecheniyu deyatel'nosti Natsional'nogo tsentra po biotekhnologii Respubliki Kazakhstan (16 November 1993), *Postanovlenie kabineta ministrov respubliki Kazakstan*, No. 1140; Sozdan Natsional'nyi tsentr po biotekhnologii (1–15 January 1994), *Nauka Kazakhstana*, No. 1(13), p. 1; *National Centre on (sic) Biotechnology Republic of Kazakhstan*, document provided to Dr Anthony Rimmington by Yurii Aleksandrovich Shapovalov, Deputy Director of the National Biotechnology Centre, Institute of Experimental Biology, Almaty, 21 April 1995.
140. Sit'ko, N., Biotekhnologicheskii proryv ne za gorami…, *Nauka Kazakhstana*, No. 4(40), 16–28 February 1995, p. 1; *Kazakhstanskaya pravda*, 8 February 1996.
141. Nesterenko, S., Kriger, V. (28 May 1993), Ot supersekretnykh vaktsin k sredstvu bor'by so SPIDom. Takov potentsial Stepnogorskogo kompleksa "Biomedpreparat", *Kazakhstanskaya pravda*, No. 122, p. 7.
142. Interview with Aleksandr Petrovich Mikhailichenko (Deputy Director *KNPK Biomedpreparat*), Stepnogorsk, 13 July 1995.
143. Ibid.
144. Meeting with representative of UK equipment distributor, Moscow, 4 June 1993.
145. UK DTI OSTEMS Mission to Kazakhstan including visit to Science Industrial Complex *Biomedpreparat*, Aksu, near Stepnogorsk, 14 July 1995.
146. Interview with general director of the National Centre for Biotechnology of the Republic of Kazakhstan, Science Industrial Complex *Biomedpreparat*, Stepnogorsk, 14 July 1995; Nesterenko, S., Kriger, V. (28 May 1993), Ot supersekretnykh vaktsin k sredstvu bor'by so SPIDom, p. 7.
147. Lepeshkin, G.N., Flagmanu farmindustrii – byt', in *Stepnogorsk. Zdes' propisany nashi serdtsa*, pp. 126–132.
148. Neiman, *Industriya mikrobov*, Znanie, Moscow, 1983; Rychkov, *Mikrobiologicheskaya promyshlennost'—meditsine, Fundamental'nye-meditsine*.

149. Interview with director of the National Centre for Biotechnology of Kazakhstan director of the Institute of Pharmaceutical Biotechnology, Moscow, 17 August 1995.
150. Lepeshkin, G.N., Flagmanu farmindustrii – byt', pp. 126–132.
151. Ibid.
152. Interview with UK distributer of medical and biopharmaceutical equipment, 21 October 1996.
153. Dobbs, M., Soviet-Era Work on Bioweapons Still Worrisome, *The Washington Post*, 12 September 2000, https://www.washingtonpost.com/archive/politics/2000/09/12/soviet-era-work-on-bioweapons-still-worrisome/f418db43-daa2-44b4-a776-eb42047ea1b8/, Accessed on 15 September 2020.
154. 18 Jul 95 International Market Insight Reports: Kazakhstan: Almaty Business Roundtable (April 7 '95), Asia Intelligence Wire, *FT Profile*, 10:35, 14 October 1997.
155. 16 May 96 ICN Pharma, Wash DC Firm in Kazakh Joint Venture, Dow Jones News Database, *FT Profile*, 10:33, 14 October 1997; Mar 97, ICN Expands its Global Business through Acquisitions, Predicasts PROMT, Med Ad News, *FT Profile*, 10:30, 14 October 1997; 01 Jun 96, ICN Forms Joint Venture, Predicasts PROMT, Applied Genetics News, *FT Profile*, 10:32, 14 October 1997; 15 May 96 US Drug Company to Convert Kazakh Bio-Weapons Plant, The Associated Press, *FT Profile*, 10:33, 14 October 1997; 14 May 96 ICN Pharmaceuticals and Allen & Associates to Convert Former Soviet Scientific Production Complex into…, Reuter Textline, PR Newswire, *FT Profile*, 10:34, 14 October 1997.
156. Dobbs, Soviet-Era Work On Bioweapons Still Worrisome.
157. Rimmington, A., Conversion of BW Facilities in Kazakhstan, p. 180 in Geissler, E., Gazsó, L., Buder, E. (Eds.), *Conversion of Former BTW Facilities*, NATO Science Series, Kluwer Academic Publishers, London, 1998 p. 180.
158. Roffey, R., Westerdahl, K.S., Conversion of former biological weapons facilities in Kazakhstan, p. 19.
159. Ibid.
160. Leitenberg, Zilinskas, with Kuhn, *The Soviet Biological Weapons Programme*, p. 281.
161. Numbers of Scientific Personnel Employed Within *Glavmikrobioprom* Research Network, Encyclopaedia of the Basic Technical-Economic Indicators of the Microbiological Industry for 1966–1985, *RGAE*, Fond 178, Opus 1, Delo 5416.
162. Dickman, S., Lithuanian biochemist builds enzyme empire, *Science*, Vol. 257, 11 September 1992, p. 1473.
163. *Institute of Biotechnology 1975–2000*, Vilnius, L.C. leidinys, 2000, p. 26.

164. Ibid.
165. Rimmington, *Who's Who*, p. 313.
166. Leitenberg, Zilinskas, with Kuhn, *The Soviet Biological Weapons Programme*, p. 164.
167. Alibek with Handelman, *Biohazard*, pp. 186–187.
168. Vorob'ev, A.A., *Dvadsatyi vek moimi glazami*, p. 268.
169. Dickman, S., Lithuanian biochemist builds enzyme empire, p. 1474.
170. Rimmington, *Who's Who*, p. 127.
171. *Institute of Biotechnology 1975–2000*, p. 27.
172. Dickman, S., Lithuanian biochemist builds enzyme empire, p. 1474.
173. Ibid.
174. Information provided by Fermentas staff to author during visit to Institute of Applied Enzymology Fermentas, Vilnius, Lithuania, 7 December 1990.
175. *Institute of Biotechnology 1975–2000*, p. 29.
176. Ibid.
177. Dickman, S., Lithuanian biochemist builds enzyme empire, p. 1474.
178. Ibid.
179. *Institute of Biotechnology 1975–2000*, p. 27.
180. Alibek with Handelman, *Biohazard*, p. 187.
181. Hoffman, *The Dead Hand*, p. 359.
182. Janulaitis, A., Naktinis, V., Biotechnology in Lithuania today, *Izvestiya Latviiskoi Akademii Nauk*, No. 12, December 1990, pp. 93–97, translated in *JPRS Report, Science & Technology, USSR: Life Sciences*, JPRS-ULS-91-013, 17 June 1991, p. 21.
183. Information provided by Fermentas staff to author during visit to Institute of Applied Enzymology Fermentas, Vilnius, Lithuania, 7 December 1990.
184. Dickman, S., Lithuanian biochemist builds enzyme empire, *Science*, p. 1474.
185. Fax message sent to the author by Arvydas Janulaitis, Institute of Biotechnology Fermentas, 21 October 1993.
186. *Institute of Biotechnology 1975–2000*, p. 31.
187. Thermo Fisher Scientific to Purchase Fermentas for $260 Million, *Genetic Engineering & Biotechnology News*, 27 May 2010, https://www.genengnews.com/topics/omics/thermo-fisher-scientific-to-purchase-fermentas-for-260m/, Accessed on 13 July 2020.
188. Personal communication to author from Derek Lennon, Chairman of Oxford Virology plc, DCL/4244R.rek, 19 March 1991, p. 1.
189. SibEnzaim, Novosibirskii oblastnoi tsentr razvitiya promyshlennosti i predprinimatel'stva, http://mspnso.ru/business/pages/view/sibenzaim, Accessed on 7 July 2020.

190. Interview with company staff on Pavilion No. 1, Stand 000 EKSO-SERVIS, Sixth International Healthcare, Medical Technology & Pharmaceuticals Exhibition Zdravookhranenie '95, Ekspotsentr, Moscow, 8 December 1995.
191. Ibid.
192. *AO Omutninskii khimicheskii zavod*, leaflet produced by the Joint Stock Company "Omutninskii khimicheskii zavod", 1995.
193. *Meditsinskii biznes*, No. 12(18), 1995, p. 1.
194. Ibid.
195. The clean rooms were originally designed for the production of 15,000,000 infusions per annum. However, the agreement with the South Korean company collapsed after the disintegration of the Soviet Union. The rooms were subsequently made ready for use by the company's personnel.
196. Promotional brochure for the Omutninsk Chemical Factory, place and date of publication unknown; *AO Omutninskii khimicheskii zavod*, leaflet produced by the Joint Stock Company Omutninskii khimicheskii zavod, 1995.
197. Interview with company staff on Pavilion No. 1, Stand 000 EKSO-SERVIS, Sixth International Healthcare, Medical Technology & Pharmaceuticals Exhibition Zdravookhranenie '95, Ekspotsentr, Moscow, 8 December 1995.
198. Bortnikov, V., Na protivorakovom napravlenii, *Vyatskii krai*, 10 January 2001.
199. *Reducing Nuclear and Biological Threats at the Source*, Hearing Before the Subcommittee on Prevention of Nuclear and Biological Attack of the Committee on Homeland Security U.S. House of Representatives, One Hundred Ninth Congress, Second Session, 22 June 2006, Serial No. 109-87, U.S. Government Printing Office, Washington, D.C., 2007, p. 17.
200. Perechen' ob'edinenii, predpriyatii i organizatsii, vkhodyashchikh v Gosudarstvennyi kontsern "Biopreparat", *Russian State Archive of the Economy* (RGAE), Fond 788, Opis 1, p. 1408.
201. Ibid.
202. Sibbiopharm: From Bioweapons to Biologics, https://www.crdfglobal.org/success-stories/sibbiopharm-bioweapons-biologics, Accessed on 27 August 2020.
203. Napali na direktora holdinga, *Novaya sibir'*, newsib.cis.ru/2004/design/2004_19/main.htm, Accessed on 25 September 2006.
204. Sibbiopharm: From Bioweapons to Biologics.
205. *The Unique U.S.-Russian Relationship in Biological Science and Biotechnology: Recent Experience and Future Directions*, Committee on

U.S.-Russian Bioengagement; Development, Security, and Cooperation; Policy and Global Affairs; National Research Council; Russian Academy of Sciences, National Academies Press, Washington (DC), 2013, https://www.ncbi.nlm.nih.gov/books/NBK201551/, Accessed on 8 September 2020.

206. Russian scientists visit MSU to receive training in fermentation pilot-plant methods, 13 August 2012, https://www.egr.msu.edu/news/2012/08/13/russian-scientists-visit-msu-receive-training-fermentation-pilot-plant-methods, Accessed on 8 September 2020.
207. Sibbiopharm modernizing biologics in Siberia, YouTube, https://www.youtube.com/watch?v=sPchzOBnuK8&list=UU_l4VuT8git9MfUkxaecd8Q&index=34, Accessed on 8 September 2020.
208. Program update: An international success story at Sibbiopharm, https://www.egr.msu.edu/news/2013/10/22/program-update-international-success-story-sibbiopharm, Accessed on 8 September 2020.
209. Russian scientists visit MSU to receive training in fermentation pilot-plant methods.
210. *Reducing Nuclear and Biological Threats at the Source*, p. 17.
211. Sibbiopharm: From Bioweapons to Biologics.
212. *The Unique U.S.-Russian Relationship in Biological Science and Biotechnology: Recent Experience and Future Directions.*
213. Sibbiopharm: From Bioweapons to Biologics.
214. Zilinskas, R.A., Mauger, P., *Biosecurity in Putin's Russia*, Lynne Rienner Publishers, Boulder, Colorado, 2018, p. 15.
215. 25–27 April 2005—Visit to Moscow, Russian Federation by the Sub-Committee on the Proliferation of Military Technology [Science & Technology], http://www.nato-pa.int/Default.asp?SHORTCUT=716, Accessed on 21 June 2006.
216. *Annual Independent Review of the UK Government's Global Health Strategy: Working with Brazil, Russia, India, China & South Africa*, Mott MacDonald, London, 2010, p. 17, https://assets.publishing.service.gov.uk/government/uploads/system/uploads/attachment_data/file/216036/dh_118808.pdf, Accessed on 16 September 2020.
217. Rimmington, *Stalin's Secret Weapon*, pp. 203–207.

CHAPTER 9

Conclusion

Characteristics of the *Biopreparat* BW Programme

The *Biopreparat* BW programme was the largest such endeavour the world has ever seen. Its network employed at least 30,000 personnel. This is quite an astonishing figure given that Leitenberg and Zilinskas estimate there were only 8000 people involved in the entire US BW programme, 5000 people employed by the Japanese in their endeavour, the British and Canadian BW programme involving probably fewer than 5000, and the Iraqi programme, fewer than 500.[1] It can therefore safely be stated that *Biopreparat*'s programme was of global significance since it clearly eclipses the numbers employed in historic BW programmes pursued by other countries. It was also one of the world's longest-running military biological efforts, having been initiated in April 1974 and terminated only in the wake of the Soviet Union's collapse in December 1991.

Biopreparat had three distinct and, at times overlapping, phases of its existence. From April 1974 through to the collapse of the USSR in December 1991, and indeed some undetermined time thereafter, it was engaged in a major offensive BW research programme and also maintained five dual-use manufacturing plants. Beginning in March 1985, with the election by the Politburo of Mikhail Gorbachev as General Secretary of the CPSU, *Biopreparat* increased in size and political importance and began also to emerge as the major player with regard to the civil

A. Rimmington, *The Soviet Union's Invisible Weapons of Mass Destruction*, https://doi.org/10.1007/978-3-030-82882-0_9

biopharmaceutical sector. By 1988, in this second phase, *Biopreparat* embraced around 40 facilities in 15 different cities and had control of domestic production of antibiotics, vaccines and other essential medicines. It was at this precise time that Kanatzhan Baizakovich Alibekov (later to become well-known as Ken Alibek) emerged as a key interlocutor between *Biopreparat* and a number of Western biotechnology companies. In the third and final phase during the 2000s, *Biopreparat* was divested of control of nearly all R&D institutes and manufacturing facilities. It assumed a highly diminished role, focused on export control and other matters and only maintained a small headquarters staff.

A key characteristic of the *Biopreparat* programme during the Soviet period was its operation of a network of mobilization production facilities which had originally been created by Khrushchev in the 1950s. This initially comprised of four, ostensibly civil manufacturing plants at Berdsk, Omutninsk, Penza and Kurgan, each of which incorporated capacity for the production of BW agents in wartime emergency. In 1970, construction of the fifth and largest mobilization plant was initiated by the Soviet Union at Stepnogorsk in Kazakhstan. Upon its creation in 1974, *Biopreparat* absorbed all of these pre-existing facilities and then proceeded to significantly expand the dual-use network with the creation of three so-called Experimental-Industrial Bases—which were pilot and industrial-scale BW facilities—at Berdsk, Omutninsk and Stepnogorsk. The latter, the Stepnogorsk Scientific Experimental-Industrial Base (*SNOPB*), was of particular importance. It is estimated to have received one billion roubles in investment and is reported to have eventually incorporated capacity for the production of around 300 metric tonnes of weaponized *B. anthracis* spores per ten-month production cycle.

Another feature of *Biopreparat* was its creation of five highly powerful R&D facilities utilizing the latest advances in genetic engineering with the aim of yielding bacterial and viral strains with wholly new and unexpected properties. In this way classical BW agents could be made even more virulent. It was believed that these new agents would be essentially undetectable, and might even be able to circumvent the provisions of the 1972 Biological and Toxin Weapons Convention. Some senior Soviet figures believed that such genetically engineered pathogens could be used to disable human targets in much the same manner as teargas and related agents.

The *Biopreparat* network recruited large numbers of personnel from the USSR Ministry of Defence's BW facilities. The key example of this phenomenon was the All-Union Scientific-Research Design Institute of

Applied Biochemistry (*VNIIbiokhimmashproekt*). It was the most pivotal *Biopreparat* institution charged with the maintenance of the dual-use mobilization plants and the creation of new, advanced R&D institutes, with a focus on the use of molecular biology for the development of novel BW agents. From its very inception, it was intimately intertwined with the USSR Ministry of Defence's Scientific-Research Technical Bureau (*NITB*) which up until this point had itself had responsibility for management of the USSR's dual-use BW facilities. Of the 16 initial core staff members at *VNIIbiokhimmashproekt* when it was created in April 1974, no less than nine were former high-level *NITB* personnel. This phenomenon of military staff occupying key managerial positions was repeated across the *Biopreparat* network.

As with other parts of the Soviet BW programme, one of the defining features of the *Biopreparat* effort were the enormous risks taken with regards to the health and safety of scientific personnel during the course of research on highly dangerous pathogens. The worst such event reported within *Biopreparat* concerned an incident at its virology facility near the Siberian city of Novosibirsk, the All-Union Scientific-Research Institute of Molecular Biology (*VNII MB*). Here, in mid-April 1988, one of its scientists, Nikolai Vasil'evich Ustinov, was, during the course of a delicate laboratory procedure, inadvertently infected with Marburg virus. He died on 30 April in terrible agony, bleeding profusely. Such incidents were not confined to *Biopreparat* and serious outbreaks were also reported at BW institutions belonging to the USSR Ministry of Defence and the USSR Ministry of Agriculture. Indeed, the release on 2 April 1979 of an anthrax aerosol from the military's Scientific-Research Institute of Bacterial Vaccine Preparations (Sverdlovsk) resulted in the deaths of at least 68 people. This was the first major indication in the West that the Soviet Union had embarked upon an offensive biological warfare programme.

Soviet Rationale for the Launch of the *Biopreparat* Programme

The Soviet rationale for the launch of the *Biopreparat* programme in April 1974 and the allocation of such immense resources to its covert BW network is complex and has been examined in considerable detail by Leitenberg and Zilinskas.[2] There is no doubt, that at that time, owing in large measure to Trofim Lysenko's political suppression of Mendelian genetics, the Soviet Union lagged well behind the West with regard to its rapid advances in molecular biology. Nixon had terminated the US offensive BW programme in 1969. However, military hawks in the Soviet leadership appear to have believed that elements of the US military biological effort continued to be maintained in secret. They were greatly concerned that the Soviet Union had been left far behind in the biological arms race.

A key individual in the decision to create the *Biopreparat* network was Academician Yurii Anatol'evich Ovchinnikov. He was the most brilliant Soviet scientist of his generation with regard to the newly emerging genetic engineering technology. He had enormous political influence and by 1974 had risen to the position of Vice President of the USSR Academy of Sciences. Later he was elected as a Candidate Member of the CPSU Central Committee and was also one of the two chief advisors to the Politburo on questions relating to science. Ovchinnikov had intimate knowledge of the warped priorities of the Soviet political system and was acutely aware that this meant that only scientific projects with a strong defence emphasis were likely to garner support from the CPSU Central Committee. Ovchinnikov convinced the lead officials, both within the Soviet microbiological industry (*Glavmikrobioprom*) and the USSR Ministry of Defence, that the creation of the covert *Biopreparat* network represented their best opportunity of securing large-scale research funding and catching up with the West in molecular biology.

The Achievements of the *Biopreparat* Programme

In a Soviet system subject to the highest levels of secrecy, it is extremely challenging to undertake any penetrative academic analysis relating to the specific military and civil achievements of the *Biopreparat* programme. Which agents, for example, were genetically engineered by *Biopreparat* with subsequent preparation of weaponization and production protocols? Did the possession of this new generation of enhanced agents strengthen the Soviet military? Were technologies which might be beneficial to the development of the civil life sciences sector spun-off from the military programme? One of the avenues open to a scholar of the Soviet Union is to attempt to identify the award of high-level Soviet state honours to participants in the *Biopreparat* programme. The Defence Department of the Central Committee of the Communist Party of the Soviet Union played a pivotal role in drafting the lists of those to be awarded key honours such as the Lenin Prize.[3] As a consequence, leading defence scientists working within the Soviet military-industrial complex on nuclear, chemical and biological programmes were showered with prizes and honours by the Soviet authorities. The citation linked to the award may identify both the scientists involved and provide an indication of the nature of the respective military achievement. Appendix D presents the results of a brief review of the literature undertaken by the author. It reveals, for example, that in 1985, Ernst Georgievich Malygin, a head of department at *Biopreparat*'s *VNII MB* (Kol'tsovo), was awarded the Lenin Prize, among the highest awards in the Soviet Union, for his work on the use of genetic engineering methods to create a microbial strain producing biologically active peptides.[4] Other recipients of the Lenin Prize within *Biopreparat* included the head of the covert network, Yurii Tikhonovich Kalinin.[5]

However, even an exhaustive search for Soviet honours bestowed upon *Biopreparat* scientists is only ever likely to provide some small inkling of the military projects in which they were involved and the accomplishments which followed. Academic scholars are therefore reliant on the very few published memoirs of weapons scientists, alongside a small pool of *Biopreparat* personnel who were interviewed by Leitenberg and Zilinskas in their monumental study of the Soviet BW programme. These sources reveal that *VNII PM*, one of the most militarized within the *Biopreparat* system, accounted for three out of the five accomplishments of the Soviet second-generation BW programme in bacteriology. In 1986, *VNII PM* researchers succeeded in genetically engineering a strain of *Bacillus anthracis* which was resistant to seven or eight antibiotics. Multi-resistant strains of *F. tularensis* (which causes tularaemia), *B. mallei* (which causes glanders) and *B. pseudomallei* (which causes melioidosis) were also created.[6] Another significant achievement of *Bonfire* concerned the insertion of virulence genes from *Bacillus cereus* into *B. anthracis.* This work was apparently completed in 1993 and raised fears that the West's existing defence anthrax vaccines would be ineffective against the new genetically engineered pathogen.[7]

Another major success attributed to *VNII PM* concerned the genetic engineering of *Legionella pneumophila* (the causative agent of Legionnaire's disease). From 1986 to 1987, an Obolensk team succeeded in developing a synthetic gene that coded for the production of a myelin peptide and to insert this gene into an expression plasmid. Eventually, *L. pneumophila* was selected for introduction of the plasmid. The goal was then to use this genetically engineered bacterium to produce a protein resembling myelin during an infection. The modified strain carrying the myelin genes was subsequently injected into guinea pigs which at first exhibited symptoms of mild pneumonia. Later the guinea pigs suffered brain damage and paralysis and then died with a mortality rate approaching 100 per cent.[8] In sharp contrast to these military successes, with regard to civil achievements in fundamental science, including molecular biology, genetics and so on, Obolensk "*was the worst—the most poorly run and the least productive—of Biopreparat's facilities*".[9]

The crucial importance of *III* (Lyubuchany) within the *Biopreparat* network is pointed to by the analysis of Leitenberg and Zilinskas who argue that the institute was responsible for two of the five major accomplishments of the Soviet second-generation BW programme in bacteriology. The first of these involved the coating of *F. tularensis* cells with Protein A to allow them to bypass the immunodefence system in humans. The large-scale production of Protein A is reported to have been achieved at the Omutninsk Chemical Factory (see above). The second is reported to have involved the genetic engineering of an F1-minus strain of *Yersinia pestis* to alter the antigenic presentation of the pathogen's surface. This would have led to delays in the detection and identification of the pathogen in any country attacked with biological weapons armed with this strain.[10] Meanwhile, *VNII MB* (Kol'tsovo) was responsible for *Biopreparat*'s main achievement with regard to virology. The institute is reported to have developed technology for the production of a dry formulation of the Marburg virus which could be dispersed in an aerosol.[11]

Alongside *Biopreparat*'s military achievements, there were also some significant accomplishments with regard to the civil sector. Perhaps the most spectacular success concerned the development of capacity for the domestic production of restriction enzymes, which are vital for genetic engineering research. In March 1975, *Biopreparat* established the All-Union Scientific-Research Institute of Applied Enzymology (*VNIIPE*) in Vilnius, Lithuania. The new facility specialized in the development and production of both highly purified restriction enzymes for molecular biology research and industrial (bulk) enzymes. In 1989, Arvydas Janulaitis, one of the very few entrepreneurs of note ever to emerge in the Soviet life sciences sector, was appointed director of *VNIIPE* and immediately took the decision to refocus the bulk of its resources on restriction enzymes. Prior to the collapse of the USSR, the facility had screened more than 25,000 microbial strains and was producing around 83 restriction enzymes and 26 DNA modification enzymes. *VNIIPE* rapidly became one of the world leaders in this area of scientific research, and by 2000, it is estimated that it had accounted for around 30 per cent of the more than 3000 restriction enzymes which had been isolated across the globe. One of the outstanding achievements claimed by the institute concerns the first discovery anywhere in the world in 1983 of the minor DNA base (N^4-methylcytosine, m^4C).[12] By September 1992, orders were being received from 15 companies based in Europe, the US and Japan, including US-based New England Biolabs. The annual income in 1991 from these sales was reported to be US$340,000.[13] Later, in the post-Soviet period, the restriction enzymes business was spun-off and eventually was acquired by Thermo Fisher Scientific in 2010 for US$260 million in cash.[14]

The success of *Biopreparat*'s restriction enzymes project does not appear to be typical of its contributions to the Soviet civil sector as a whole. It is notable that the commercialization of its capabilities in this area was achieved by scientists in Lithuania where the desire to break out of the Soviet strangle-hold restricting scientific innovation was at its strongest. However, there were few similar success stories in other parts of the network. *Biopreparat*'s creaking bureaucratic management structures were totally unsuited to innovative R&D and there was a singular failure to use the newly available genetic engineering technologies to create new medicines for the Soviet health service. It is evident that the life sciences sector in the USSR fell far behind that of the West, which was developing and commercially producing a range of recombinant products, such as human insulin.

To sum up then, after astronomical expense, *Biopreparat* did succeed in developing a small number of second-generation BW agents which were turned over to the USSR Ministry of Defence for the ultimate stages of weaponization. Whether these agents made any substantive enhancement of Soviet capabilities to wage war must remain very much open to question. Meanwhile, the anticipated civil spin-offs from the *Biopreparat* programme were largely absent and there was no delivery pipeline of novel vaccines and medicines for the Soviet health service. The detrimental economic impact of the inordinately wasteful pursuit of the secret programme is unknown, but it undoubtedly contributed to the eventual collapse of the Soviet Union—the country unable to sustain the ruinous cost of such grandiose military endeavours.

BIOPREPARAT'S LEGACY

There is evidence that key parts of the *Biopreparat* R&D network have been demilitarized. In their 2018 study, Zilinskas and Mauger, note that the Scientific-Research Institute of Highly Pure Biopreparations (St Petersburg) and the Institute of Engineering Immunology (Lyubuchany) "*appear to conduct strictly civilian research under transparent conditions*".[15] The authors indicate that at some point between 1991 and 1992, all secret work at the *Vektor* virology centre in Kol'tsovo was also terminated. However, the situation at the State Scientific Centre of Applied Microbiology and Biotechnology (Obolensk), which remains functional, appears less clear-cut and since around 2009 has been mostly closed to foreigners.

There have been similar moves to eliminate or reorientate *Biopreparat*'s former network of dual-use mobilization facilities to civil commercial objectives. In a major non-proliferation project funded by the US, the Scientific Experimental-Industrial Base (*SNOPB*) in Stepnogorsk has been dismantled. This project undertaken in collaboration with the government of Kazakhstan represents an achievement of global significance, eliminating the most advanced and sophisticated BW production complex the world has ever seen. A similar success was achieved by the US and Russia with regard to the *Sibbiofarm* Production Association which incorporates the Berdsk biologics plant. The latter was the largest dual-use biological production facility in Russia. Sometime around 2007 or thereafter, the dual-use BW facilities at Berdsk, which had been in existence for more than half a century, were removed and destroyed.

Russia appears to have reorientated *Biopreparat*'s remaining dual-use mobilization facilities to civil commercial objectives. Following the collapse of the USSR, the Omutninsk Chemical Factory was reorganized as two joint stock companies with production initially focused on the manufacture of disposable syringes. Another major facility at the Vostochnyi site during the Soviet period was the Omutninsk Scientific Experimental-Industrial Base (*ONOPB*). It too was transformed into a joint stock company and has been engaged in the production of antitumour drug substances, including bleomycin. Little has been published about *Biopreparat*'s former mobilization facilities in Penza and Kurgan, but these appear to be focused now on the production of antibiotics for the Russian health service.

What of the *Biopreparat* scientists themselves? Only a proportion of these would have possessed expertise with regard to BW agents. Those personnel working in specialized institutes or design bureaus, such as the staff at *VNIIbiokhimmashproekt* concerned with the design and optimization of BW facilities, or those employed in facilities concerned in the design of specialized filling equipment or delivery systems, would also be of particular proliferation concern. According to Zilinskas and Mauger writing in 2018, of those *Biopreparat* scientists with the highest security clearance, most are now retired. Of the remainder, many were hired by the three military biological facilities belonging to the Russian Ministry of Defence.[16] It is not only *Biopreparat* personnel which have been absorbed by the facilities belonging to the Russian military. *Biopreparat*'s pathogenic seed strains—including those of genetically engineered pathogens—along with detailed documentation relating to methods of their weaponization, are believed to have been transferred to secret repositories within the military BW network.[17]

Given the above, the West's focus of attention with regard to Russia's BW activity has inevitably shifted from *Biopreparat* to the three major military biological institutes—the Russian Ministry of Defence's 48th Central Scientific-Research Institute (Kirov) and its branch facilities in Ekaterinburg and Sergiev Posad. These institutes all remain operational and closed to foreigners and Russia has been repeatedly challenged with regard to a lack of transparency concerning the nature of their activities. It is highly likely that the fruit of *Biopreparat*'s labours over nearly two decades, the seed strains and "recipes" detailing weaponization and production protocols, are retained within these secretive military facilities.[18] Such is Western mistrust of the network that, in August 2020, the US Commerce Department's Bureau of Industry and Security (BIS) imposed "blacklist" restrictions on the three military biological institutes for their alleged involvement in biological weapons research.[19] On 2 March 2021, additional US sanctions were imposed on the institutes in relation to the poisoning of Russian opposition leader Aleksei Navalny.[20] There do therefore remain major uncertainties with regard to the fate of seed strains and associated weapons technologies developed by *Biopreparat*. These sit alongside a wider Western mistrust concerning the nature and goals of the Russian military's current biological defence programmes.

The heightened international tensions which exist between the US and its allies and the Russian Federation, encompassing a range of issues, are clearly not conducive to bilateral and multilateral engagements in the life sciences. Nevertheless, Gronvall and Bland argue in their recent paper that attempts to engage with Russian scientists and institutions in this area are important. They maintain that "*programmes that bring together scientists from Russia with US scientists should be a priority, as it can be helpful to diplomacy if scientists work together on uncontroversial issues regarding biological sciences research advances or norms for the conduct of research*".[21] Despite, or perhaps because of, the very poor state of the West's relations with Russia, it is vitally important that such engagement should continue to be pursued.

Notes

1. Leitenberg, Zilinskas with Kuhn, *The Soviet Biological Weapons Programme*, p. 700.
2. Ibid., pp. 51–78.
3. Sagdeev, *The Making of a Soviet Scientist.*
4. Prizes, http://vector.sibinfo.ru/DesktopDefault.aspx?lcid=25&tabid=21&tabindex=1, Accessed on 6 May 2004.
5. Kalinin Yurii Tikhonovich, Mezhdunarodnyi ob"edinenyi biograficheskii tsentr, http://www.biograph.ru/index.php/whoiswho/11-medicine/474-kalinin-yt, Accessed on 13 October 2020.
6. Leitenberg, Zilinskas, with Kuhn, *The Soviet Biological Weapons Programme*, pp. 702–703.
7. Miller, Engelberg, Broad, *Germs*, p. 208.
8. Hoffman, *The Dead Hand*, p. 299.
9. Domaradskij, I., Orent, W., *Biowarrior*, p. 246.
10. Leitenberg, Zilinskas, with Kuhn, *The Soviet Biological Weapons Programme*, pp. 702–703.
11. Ibid., pp. 218–219.
12. *Institute of Biotechnology 1975–2000*, p. 27.
13. Dickman, S., Lithuanian biochemist builds enzyme empire, p. 1474.
14. Thermo Fisher Scientific to Purchase Fermentas for $260 Million, *Genetic Engineering & Biotechnology News*, 27 May 2010, https://www.genengnews.com/topics/omics/thermo-fisher-scientific-to-purchase-fermentas-for-260m/, Accessed on 13 July 2020.
15. *Biosecurity in Putin's Russia*, pp. 15–16, 23.
16. Ibid., p. 18.
17. Ibid., pp. 18–19.
18. Ibid., p. 23.

19. Addition of Entities to the Entity List, and Revision of Entries on the Entity List, A Rule by the Industry and Security Bureau, *Federal Register*, 27/08/2020, https://www.federalregister.gov/documents/2020/08/27/2020-18909/addition-of-entities-to-the-entity-list-and-revision-of-entries-on-the-entity-list, Accessed on 9 March 2021.
20. U.S. Sanctions and Other Measures Imposed on Russia in Response to Russia's Use of Chemical Weapons, Fact Sheet, Office of the Spokesperson, 2 March 2021, Chemical and Biological Weapons Act Sanctions, https://ru.usembassy.gov/u-s-sanctions-and-other-measures-imposed-on-russia-in-response-to-russias-use-of-chemical-weapons/, Accessed on 9 March 2021.
21. Gronvall, G.K., Bland, B., Life-science research and biosecurity concerns in the Russian Federation, *The Nonproliferation Review*, 5 February 2021, https://www.tandfonline.com/doi/pdf/10.1080/10736700.2020.1866323, p. 9, Accessed on 10 March 2021.

Appendix A: Soviet and Russian Abbreviations and Acronyms

BelNIIEM	Belorussian Scientific-Research Institute of Epidemiology and Microbiology (*Belorusskii nauchno-issledovatel'skii institut epidemiologii i mikrobiologii*)
BNOPB	Berdsk Scientific Experimental-Industrial Base (*Berdskaya nauchnaya opytno-promyshlen naya baza*)
FBUN GNTs PMB	Federal Budgetary Science Institution "State Scientific Centre of Applied Microbiology and Biotechnology" (*Federal'noe byudzhetnoe uchrezhdenie nauki "Gosudarstvennyi nauchnyi tsentr prikladnoi mikrobiologii i biotekhnologii"*)
FBUN GNTs VB Vektor	Federal Budgetary Science Institution "State Scientific Centre of Virology and Biotechnology Vektor" (*Federal'noe byudzhetnoe uchrezhdenie nauki "Gosudarstvennyi nauchnyi tsentr virusologii i biotekhnologii Vektor"*)
Glavmikrobioprom	Main Administration of the Microbiological Industry under the USSR Council of Ministers (*Glavnoe upravlenie mikrobiologicheskoi promyshlennosti pri Sovete Ministrov SSSR*)

A. Rimmington, *The Soviet Union's Invisible Weapons of Mass Destruction*, https://doi.org/10.1007/978-3-030-82882-0

III	Institute of Engineering Immunology (*Institut inzhenernoi immunologii*)
KNPK Biomedpreparat	Kazakh Science Production Complex *Biomedpreparat* (*Kazakhstanskii nauchno-promyshlennyi kompleks "Biomedpreparat"*)
LOBP	Lotoshino Experimental Biotechnology Enterprise (*Lotoshinskoe opytnoe biotekhnologicheskie predpriyatie*)
Minmedbioprom	USSR Ministry of the Medical and Microbiological Industry (*Ministerstvo meditsinskoi i mikrobiologicheskoi promyshlennosti SSSR*)
Minmedprom	USSR Ministry of the Medical Industry (*Ministerstvo meditsinskoi promyshlennosti SSSR*)
Minsredmash	USSR Ministry of Medium Machine Building (*Ministerstvo srednego mashinostroeniya SSSR*)
Minzdravmedprom	Ministry of Health and the Medical Industry of the Russian Federation (*Ministerstvo zdravookhraneniya i meditsinskoi promyshlennosti Rossiiskoi Federatsii*)
MNTS	Interdepartmental Scientific-Technical Council for Molecular Biology & Genetics (*Mezhdomstvennyi nauchno-tekhnicheskii sovet po molekulyarnoi biologii i genetike*)
NIKTI BAV	Scientific-Research Design Institute of Biologically Active Substances (*Nauchno-issledovatel'skii konstruktorsko-tekhnologicheskii institut biologicheski aktivnykh veshchestv*)
NITB	Scientific-Research Technical Bureau (*Nauchno-issledovatel'skoe tekhnicheskoe byuro*)
NOPB	Scientific Experimental-Industrial Base (*Nauchnaya opytno-promyshlennaya baza*)
NPO Biomash	Science Production Association *Biomash* (*Nauchno-proizvodstvennoe ob"edinenie Biomash*)
NPO Promavtomatika	Science Production Association *Promavtomtika* (*Nauchno-proizvodstvennoe ob"edinenie Promavtomatika*)

NPO Vektor	Science Production Association *Vektor* (*Nauchno-proizvodstvennoe ob"edinenie Vektor*)
NTsB	Kazakhstan National Centre for Biotechnology (*Natsional'nyi tsentr po biotekhnologii Respubliki Kazakhstan*)
OAO Biopreparat	Open Joint Stock Company *Biopreparat* (*Aktsionernoe obshchestvo otkrytogo tipa Biopreparat*)
OAO Biopreparat-Tsentr	Open Joint Stock Company *Biopreparat-Tsentr* (*Aktsionernoe obshchestvo otkrytogo tipa Biopreparat-Tsentr*)
OKBA	Experimental Design Bureau of Automation Equipment (*Opytno-konstruktorskoe byuro sredstv avtomatizatsii*)
OKB TBM	Experimental Design Bureau of Precision Biological Engineering (*Opytno-konstruktorskoe byuro tonkogo biologicheskogo mashinostroeniya*)
OKhZ	Omutninsk Chemical Factory (*Omutninskii khimicheskoi zavod*)
ONOPB	Omutninsk Scientific Experimental-Industrial Base (*Omutninskaya nauchnaya opytno-promyshlennaya baza*)
OOO PO Sibbiofarm	*Sibbiofarm* Production Association (*OOO Proizvodstvennoe Ob"edinenie Sibbiofarm*)
RAO Biopreparat	Russian Joint Stock Company *Biopreparat* (*Russiiskoe aktsionernoe obshchestvo Biopreparat*)
Rospotrebnadzor	Federal Service for Supervision in the Sphere of Protection of Consumer Rights and Wellbeing of the Person (*Federal'naya sluzhba po nadzoru v sfere zashchity prav potrebitelei i blagopoluchiya cheloveka*)
SKTB BAV	Special Design Bureau of Biologically Active Substances (*Spetsial'noe konstruktorsko-tekhnologicheskoe byuru biologicheski aktivnykh veshchestv*)

SNOPB	Stepnogorsk Scientific Experimental-Industrial Base (*Stepnogorskaya nauchnaya opytno-promyshlennaya baza*)
SO AN SSSR	Siberian Branch of the USSR Academy of Sciences (*Sibirskoe otdelenie Akademii Nauk SSSR*)
Sovnarkhoz	Regional Economic Council (*Sovet narodnogo khozyaistva*)
VNIIbakpreparat	All-Union Scientific-Research Institute of Microbial Pesticides and Bacterial Preparations (*Vsesoyuznyi nauchno-issledovatel'skii institut mikrobiologicheskikh sredstv zashchity rastenii i bakpreparatov*)
VNIIbiokhimmashproekt	All-Union Scientific-Research Design Institute of Applied Biochemistry (*Vsesoyuznyi nauchno-issledovatel'skii proektno-konstruktorskii institut prikladnoi biokhimii*)
VNIIbiotekhnologii	All-Union Scientific-Research Institute of Biotechnology (*Vsesoyuznyi nauchno-issledovatel'skii institut biotekhnologii*)
VNII BP	All-Union Scientific-Research Institute of Biological Instrument Making (*Vsesoyuznyi nauchno-issledovatel'skii institut biologicheskogo priborostroeniya*)
VNIIgenetika	All-Union Scientific-Research Institute of Genetics and Selection of Industrial Microorganisms (*Vsesoyuznyi nauchno-issledovatel'skii institut genetiki i selektsii promyshlennykh mikroorganizmov*)
VNII MB	All-Union Scientific-Research Institute of Molecular Biology (*Vsesoyuznyi nauchno-issledovatel'skii institut molekulyarnoi biologii*)
VNIIOChB	All-Union Scientific-Research Institute of Highly Pure Biopreparations—(*Vsesoyuznyi nauchno-issledovatel'skii institut osobo chistykh biopreparatov*)
VNIIPE	All-Union Scientific-Research Institute of Applied Enzymology (*Vsesoyuznyi nauchno-issledovatel'skii institut prikladnoi enzimologii*)

VNII PM	All-Union Scientific-Research Institute of Applied Microbiology (*Vsesoyuznyi nauchno-issledovatel'skii institut prikladnoi mikrobiologii*)
VNIIsintezbelok	All-Union Scientific-Research Institute of the Biosynthesis of Protein Substances (*Vsesoyuznyi nauchno-issledovatel'skii institut biosinteza belkovykh veshchestv*)
VNPO Biopreparat	All-Union Science Production Association *Biopreparat* (*Vsesuyuznoe nauchno-proizvodstvennoe ob"edinennie Biopreparat*)
VPK	Military-Industrial Commission (*Voenno-promyshlennaya komissiya*)

Appendix B: Composition of the Interbranch Scientific and Technical Council for Molecular Biology and Genetics (*MNTS*)

Chair: Academician Viktor Mikhailovich Zhdanov (1971–1975); Vasilii Dmitrievich Belyaev (1975–1979); Rotislav S. Rychkov (1979–1985); Valerii Alekseevich Bykov (1985–)

Secretariat (Administrative Department) Chair: Igor' Valerianovich Domaradskii (1973–1982); Major General Dmitrii Vladimirovich Vinogradov-Volzhinskii (1982–)

Deputy Chair: Colonel Lev Aleksandrovich Klyucherev (1973–)

Members

Representing *Biopreparat*: Major General Vsevolod Ivanovich Ogarkov (1973–1979); Lieutenant General Yurii Tikhonovich Kalinin (1979–); Kanatzhan Baizakovich Alibekov

Representing USSR Ministry of Agriculture: Major General Mikhail Aleksandrovich Khudyakov (~1975–1984); Petr Ivanovich Morozov

Representing USSR Ministry of Defence's Fifteenth Administration: Major General Vladimir Andreevich Lebedinskii (1971–); Colonel General Efim Ivanovich Smirnov (~1975–1987)

Representing USSR Academy of Sciences: Yurii Anatol'evich Ovchinnikov (1971–1988); Georgii Konstantinovich Skryabin (1971–); Aleksandr Aleksandrovich Baev (1971–); Andrei Dareevich Mirzabekov and Rem Viktorovich Petrov

A. Rimmington, *The Soviet Union's Invisible Weapons of Mass Destruction*, https://doi.org/10.1007/978-3-030-82882-0

Representing USSR Ministry of Health: Lieutenant General (Reserves) Avetik Ignat'evich Burnazyan (1971–)

Representing *Glavmikrobioprom*: Vasilii Dmitrievich Belyaev (1971–1975); Anatolii Aleksandrovich Skladnev (1973–); Sos Isaakovich Alikhanyan (1971–)

Appendix C: List of Associations, Enterprises and Organizations Included Within the State Concern *Biopreparat*, 4 April 1991[1]

1. Science Production Association *Biomash*

Comprising

All-Union Scientific-Research Design Institute of Applied Biochemistry (Moscow)

State Design-Investigation Institute of Enterprises of the Biological Industry (Moscow)

Experimental-Construction Bureau of Control and Automation Instruments with its Experimental Factory (Yoshkar-Ola, Mari ASSR)

2. Science Production Association *Biosintez*

Comprising

All-Union Scientific-Research Institute of Applied Microbiology (Obolensk, Moscow *oblast'*)

Institute of Immunology (Lyubuchany settlement, Moscow *oblast'*)

Lotoshino Experimental Biotechnological Enterprise (Lotoshino, Moscow *oblast'*)

3. Science Production Association *Biotekhnika*

Comprising

All-Union Scientific-Research Biotechnical Institute (Moscow)

Grozny Branch of the All-Union Scientific-Research Biotechnical Institute (Grozny, Checheno-Ingush ASSR)

Moscow Experimental Factory of the All-Union Scientific-Research Biotechnical Institute (Moscow)

A. Rimmington, *The Soviet Union's Invisible Weapons of Mass Destruction*, https://doi.org/10.1007/978-3-030-82882-0

Efremov Experimental-Mechanical Factory (Efremov, Tula *oblast'*, RSFSR)

4. Science Production Association *Vaktsiny*

Comprising

Tashkent Scientific-Research Institute of Vaccines and Sera (Tashkent, Uzbek SSR)

Enterprise for the Production of Bacterial Preparations (Tashkent, Uzbek SSR)

Laboratory Animals Nursery (Yangibazar settlement, Tashkent *oblast'*, Uzbek SSR)

Namangan Factory of Medical Items (Namangan, Uzbek SSR)

5. Science Production Association *Vektor*

Comprising

All-Union Scientific-Research Institute of Molecular Biology (Kol'tsovo settlement, Novosibirsk *oblast'*, RSFSR)

Scientific-Research Design-Technological Institute of Biologically Active Substances (Berdsk, Novosibirsk *oblast'*, RSFSR)

Experimental-Production Agricultural Enterprise (Morozovo settlement, Novosibirsk *oblast'*, RSFSR)

6. Science Production Association *Medbiospetstekhoborudovanie*

Comprising

Mariupol' Factory of Technological Equipment (Mariupol', Donetsk *oblast'*, Ukrainian SSR)

All-Union Scientific-Research Experimental-Design Institute of Special Technological Equipment and Rigging (Mariupol', Donetsk *oblast'*, Ukrainian SSR)

Experimental Production Plant of the All-Union Scientific-Research Experimental-Design Institute of Special Technological Equipment and Rigging (Mariupol', Donetsk *oblast'*, Ukrainian SSR)

7. Science Production Association *Farmpribor*

Comprising

All-Union Scientific-Research Institute of Highly Pure Biopreparations (Leningrad, RSFSR)

Experimental-Design Bureau of Fine Biological Engineering (Kirishi, Leningrad *oblast'*, RSFSR)

8. Science Production Association *Ferment*

Comprising

Kaunas Factory of Endocrine Preparations (Kaunas, Lithuanian SSR)

All-Union Scientific-Research Institute of Applied Enzymology (Vilnius, Lithuanian SSR)

Vilnius Experimental-Industrial Factory of Enzyme Preparations (Vilnius, Lithuanian SSR)

Kedainyai Biochemical Factory (Kedainyai, Lithuanian SSR)

Production-Experimental Factory *Sanitas* (Kaunas, Lithuanian SSR)

9. Production Association *Belmedpreparaty*

Minsk Factory of Medical Preparations (Minsk, Byelorussian SSR)

Minsk Factory of Endocrine Preparations (Minsk, Byelorussian SSR)

10. Production Association *Vostok*

Omutninsk Chemical Factory (Omutninsk, Kirov *oblast'*, RSFSR)

Omutninsk Scientific Experimental-Industrial Base (Omutninsk, Kirov *oblast'*, RSFSR)

11. Production Association *Progress*

Chemical Factory *Progress* (Stepnogorsk, Tselinograd *oblast'*, Kazakh SSR)

Stepnogorsk Scientific Experimental-Industrial Base (Stepnogorsk, Tselinograd *oblast'*, Kazakh SSR)

12. Production Association *Sibbiofarm*

Berdsk Factory of Biological Preparations (Berdsk, Novosibirsk *oblast'*, RSFSR)

Small State Agricultural Enterprise *Druzhba* (Nizhnyaya Kamenka, Novosibirsk *oblast'*, RSFSR)

13. All-Union Scientific-Research Institute of Biological Instrument Making (Moscow)

14. All-Union Scientific-Research Institute of Medical Polymers (Moscow)

15. Leningrad Scientific-Research Institute of Vaccines and Sera (Leningrad, RSFSR)

16. Krasnoyarsk 60 Years of the USSR Combine of Medical Preparations (Krasnoyarsk, RSFSR)

17. Kurgan *Sintez* Factory of Medical Preparations and Items (Kurgan, RSFSR)

18. Penza *Biosintez* Combine of Medical Preparations (Penza, RSFSR)

19. Saransk Combine of Medical Preparations *Biokhimik* (Saransk, Republic of Mordovia)

20. Vyshnii Volochek Factory of Enzyme Preparations (Kalinin *oblast'*, RSFSR)

21. Zugdidi Factory of Medical Items (Georgian SSR)
22. Moscow Endocrine Factory (Moscow)
23. Novosibirsk Factory of Medical Preparations (Novosibirsk, RSFSR)
24. *Biosnabkomplekt* Enterprise (Obolensk, Moscow *oblast'*)
25. Automobile Enterprise (Moscow)
26. Management of the Enterprise Being Constructed (Kalyazin, Kalinin *oblast'*, RSFSR)
27. Management of the Enterprise Being Constructed (Yalta, Crimea, Ukrainian SSR)

Appendix D: Award of Soviet State Honours to *Glavmikrobioprom* and *Biopreparat* Personnel

Year	*Prize*	*Name*	*R&D*
1960	Order of Lenin	Vasilii Dmitrievich Belyaev (head of *Glavmikrobioprom*)	For organizing the development and production of the nerve gas Sarin[2]
1980	USSR State Prize for Science & Technology	Anatolii Andreevich Vorob'ev (first deputy director of *Biopreparat*)	For his work on the development of vaccines and mass methods of vaccination[3]
1981	USSR Council of Ministers' Prize	Boris Vasil'evich Prilepskii (Berdsk Factory of Biological Preparations)	Unknown[4]
1981	USSR Council of Ministers' Prize	Vladimir Georgievich Debabov (director, *Glavmikrobioprom*'s *VNIIgenetika*)	Development of a national collection of restriction enzymes for genetic engineering research in the Soviet Union[5]
1981	Order of the Red Banner of Labour	Yurii Tikhonovich Kalinin (head of *Biopreparat*)	Unknown[6]
1981	Order of the Red Banner of Labour	Lev Stepanovich Sandakhchiev (director, Science Production Association *Vektor*)	Unknown[7]

(*continued*)

A. Rimmington, *The Soviet Union's Invisible Weapons of Mass Destruction*, https://doi.org/10.1007/978-3-030-82882-0

Year	*Prize*	*Name*	*R&D*
1982	Order of the Friendship of Peoples	Igor' Valerianovich Domaradskii (All-Union Scientific-Research Institute of Applied Microbiology)	For "*services in the development of molecular biology and genetics*"[8]
1984	Lenin Prize	Vladimir Georgievich Debabov (director, *Glavmikrobioprom's VNIIgenetika*)	For the development of a strain of *Pseudomonas putida* for production of human alpha interferon-2[9]
1984	Lenin Prize	Yurii Tikhonovich Kalinin (head of *Biopreparat*)	For the development and introduction of genetic engineering methods and their employment to produce biologically active substances[10]
1984	Lenin Prize	Ernst Georgievich Malygin (All-Union Scientific-Research Institute of Molecular Biology)	For using genetic engineering methods to create a microbial strain producing biologically active peptides[11]
1984	Lenin Prize	Viktor Isaakovich Marchenko (Institute of Immunology)	For the development and introduction of genetic engineering methods and their employment to produce biologically active substances[12]
1984	USSR State Prize	Viktor Fedorovich Panfilov (chief technologist, Fermentation Workshop, *SNOPB*, Stepnogorsk)	Unknown[13]
1985	USSR State Prize	Lev Stepanovich Sandakhchiev (director, Science Production Association *Vektor*)	For the development of unique methods of analysis of high-molecular compounds[14]
1985	USSR State Prize	Fedor Nikitovich Chegodaev (chief engineer at the Berdsk Factory of Biological Preparations)	Unknown[15]
1985	Order of the Red Badge of Honour	Vladimir Vasileevich Samukov (Laboratory of Peptide Chemistry at All-Union Scientific-Research Institute of Molecular Biology)	Unknown[16]

(*continued*)

Year	*Prize*	*Name*	*R&D*
1985	Order of Lenin	Lev Stepanovich Sandakhchiev (director, Science Production Association *Vektor*)	Unknown[17]
Unknown	Order of the Red Banner of Labour	Valerii Alekseevich Bykov (head of *Glavmikrobioprom*, Minister of *Minmedbioprom*)	Unknown[18]
Unknown	Order of the Red Banner of Labour	Boris Vasil'evich Prilepskii (Berdsk Factory of Biological Preparations)	Unknown[19]
Unknown	Order of the Red Banner of Labour	Anatolii Petrovich Sadovskii (Science Production Association *Vektor*)	Unknown[20]
Unknown	Order of the Red Banner of Labour	Nikolai Nikolaevich Urakov (director of *VNII PM*)	Unknown[21]
Unknown	Hero of Socialist Labour	Vladimir Arkad'evich Valov (director of Omutninsk Chemical Factory)	Industrial production of influenza vaccine[22]
Unknown	Order of the Red Banner of Labour	Stanislav Konstantinovich Vasilenko	Unknown[23]
Unknown	Order of the Red Banner of Labour	Konstantin Ivanovich Volkovoi (deputy director *Biopreparat*, head of department at All-Union Scientific-Research Institute of Applied Microbiology)	Unknown[24]

Notes

1. Perechen' ob"edinenii, predpriyatii i organizatsii, vkhodyashchikh v Gosudarstvennyi kontsern "Biopreparat", *RGAE*, Fond 788, Opis 1, p. 1408.
2. Laureaty Leninskoi premii, Vikepediya, https://ru.wikipedia.org/wiki/%D0%9B%D0%B0%D1%83%D1%80%D0%B5%D0%B0%D1%82%D1%8B_%D0%9B%D0%B5%D0%BD%D0%B8%D0%BD%D1%81%D0%BA%D0%BE%D0%B9_%D0%BF%D1%80%D0%B5%D0%BC%D0%B8%D0%B8#1960, Accessed on 13 October 2020.
3. Vorob'ev, *Ne podvodya cherty*, p. 411.
4. Rimmington, *Who's Who*, p. 233.
5. Ibid., pp. 121–123.

6. Kalinin Yurii Tikhonovich, Mezhdunarodnyi ob"edinenyi biograficheskii tsentr, http://www.biograph.ru/index.php/whoiswho/11-medicine/474-kalinin-yt, Accessed on 13 October 2020.
7. Rimmington, *Who's Who*, pp. 252–253.
8. Domaradskij, Orent, *Biowarrior*, p. 232.
9. Rimmington, *Who's Who*, pp. 121–123.
10. Kalinin Yurii Tikhonovich, Mezhdunarodnyi ob"edinenyi biograficheskii tsentr.
11. Rimmington, *Who's Who*, pp. 179–180.
12. Fedorov, *Sovetskoe biologicheskoe oruzhie.*
13. Telezhinskaya, N., *Stepnogorsk. Zavod Progress 70-80-e gody*, Personal blog, natalya-telezhinskaya.ru › 2020/04/05, Accessed on 1 October 2020.
14. Rimmington, *Who's Who*, pp. 252–253.
15. Ibid., p. 62.
16. Ibid., p. 252.
17. Ibid., pp. 252–252.
18. Ibid., pp. 56–57.
19. Ibid., p. 233.
20. Ibid., pp. 249–250.
21. Ibid., pp. 299–300.
22. Vorob'ev, *Dvadtsatyi vek moimi glazami*, pp. 210–212.
23. Ibid., p. 306.
24. Ibid., p. 312.

Index

A. Rimmington, *The Soviet Union's Invisible Weapons of Mass Destruction*, https://doi.org/10.1007/978-3-030-82882-0

The manufacturer's authorised representative in the EU is Springer Nature Customer Service Centre GmbH, Europaplatz 3, 69115 Heidelberg, Germany. If you have any concerns regarding our products, please contact ProductSafety@springernature.com

Printed and bound by CPI Group (UK) Ltd, Croydon, CR0 4YY

30/09/2026

02225702-0010